REVISE EDEXCEL GCSE (9–1)
Mathematics
Higher

REVISION GUIDE

Series Consultant: Harry Smith

Author: Harry Smith

A note from the publisher

In order to ensure that this resource offers high-quality support for the associated Pearson qualification, it has been through a review process by the awarding body. This process confirms that this resource fully covers the teaching and learning content of the specification or part of a specification at which it is aimed. It also confirms that it demonstrates an appropriate balance between the development of subject skills, knowledge and understanding, in addition to preparation for assessment.

Endorsement does not cover any guidance on assessment activities or processes (e.g. practice questions or advice on how to answer assessment questions), included in the resource nor does it prescribe any particular approach to the teaching or delivery of a related course.

While the publishers have made every attempt to ensure that advice on the qualification and its assessment is accurate, the official specification and associated assessment guidance materials are the only authoritative source of information and should always be referred to for definitive guidance.

Pearson examiners have not contributed to any sections in this resource relevant to examination papers for which they have responsibility.

Examiners will not use endorsed resources as a source of material for any assessment set by Pearson.

Endorsement of a resource does not mean that the resource is required to achieve this Pearson qualification, nor does it mean that it is the only suitable material available to support the qualification, and any resource lists produced by the awarding body shall include this and other appropriate resources.

Question difficulty

Look at this scale next to each exam-style question. It tells you how difficult the question is.

Worked solution video

Worked solution videos

Some of the questions in this Revision Guide have worked solution videos. Use your phone, tablet or webcam to scan the QR code to watch the video.

For the full range of Pearson revision titles across GCSE, AS/A Level and BTEC visit:

www.pearsonschools.co.uk/revise

ALWAYS LEARNING

PEARSON

Contents

A small bit of small print

Edexcel publishes Sample Assessment Material and the Specification on its website. This is the official content and this book should be used in conjunction with it. The questions in *Now try this* have been written to help you practise every topic in the book. Remember: the real exam questions may not look like this.

Factors and primes

The **factors** of a number are any numbers that divide into it exactly. A **prime number** has exactly two factors. The prime numbers are 2, 3, 5, 7, 11, 13, 17, 19 and so on.

Prime factors

If a number is a factor of another number **and** it is a prime number then it is called a **prime factor**. You use a factor tree to find prime factors.

Remember to circle the prime factors as you go along. The order doesn't matter.

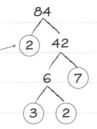

$$84 = 2 \times 2 \times 3 \times 7$$
$$= 2^2 \times 3 \times 7$$

Remember to put in the multiplication signs. This is called a **product** of **prime factors**.

The highest common factor (HCF) of two numbers is the **highest number** that is a **factor** of both numbers.

The lowest common multiple (LCM) of two numbers is the **lowest number** that is a **multiple** of both numbers.

Worked example

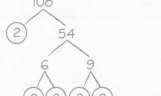

(a) Write 108 as the product of its prime factors. Give your answer in index form. **(3 marks)**

$$108 = 2 \times 2 \times 3 \times 3 \times 3 = 2^2 \times 3^3$$

(b) Work out the highest common factor (HCF) of 108 and 24. **(2 marks)**

$$108 = ②\times②\times③\times 3 \times 3$$
$$24 = ②\times②\times 2 \times③$$
HCF is $2 \times 2 \times 3 = 12$

(c) Work out the lowest common multiple (LCM) of 108 and 24. **(2 marks)**

LCM $= 12 \times 3 \times 3 \times 2 = 216$

Draw a factor tree. Continue until every branch ends with a prime number. This question asks you to write your answer in **index form**. This means you need to use **powers** to say how many times each prime number occurs in the product.

Check it!
$2^2 \times 3^3 = 4 \times 27 = 108$ ✓

To find the HCF circle all the prime numbers which are **common** to both products of prime factors. 2 appears twice in both products so you have to circle it twice. Multiply the circled numbers together to find the HCF.

To find the LCM multiply the HCF by the numbers in both products that were not circled in part (b).

You can use numbers given in index form directly:
HCF: Choose the **lowest** power of each prime
LCM: Choose the **highest** power of each prime
For example, HCF $= 2^0 \times 3^2 \times 5^0 \times 7^1 = 3^2 \times 7$

Now try this

This scale tells you how hard the question is.

1 (a) Express 980 as a product of its prime factors. **(3 marks)**

  (b) Find the highest common factor (HCF) of 980 and 56. **(2 marks)**

2 $X = 2 \times 3^5 \times 7^2$ $Y = 3^2 \times 5 \times 7$

  (a) Find the highest common factor (HCF) of X and Y. **(2 marks)**

  (b) Find the lowest common multiple (LCM) of X and Y. **(2 marks)**

Indices 1

The index laws tell you how to work with **powers** of numbers.

1 Index laws

Indices include square roots, cube roots and powers.

You can use the index laws to simplify powers and roots.

$a^m \times a^n = a^{m+n}$

$4^3 \times 4^7 = 4^{3+7} = 4^{10}$

$\dfrac{a^m}{a^n} = a^{m-n}$

$12^8 \div 12^3 = 12^{8-3} = 12^5$

$(a^m)^n = a^{mn}$

$(7^3)^5 = 7^{3 \times 5} = 7^{15}$

2 Cube root

The cube root of a positive number is positive.

$4 \times 4 \times 4 = 64$

$4^3 = 64$

$\sqrt[3]{64} = 4$

The cube root of a negative number is negative.

$-4 \times -4 \times -4 = -64$

$(-4)^3 = -64$

$\sqrt[3]{-64} = -4$

3 Powers of 0 and 1

Anything raised to the power 0 is equal to 1.

$6^0 = 1$ $1^0 = 1$ $7223^0 = 1$ $(-5)^0 = 1$

Anything raised to the power 1 is equal to itself.

$8^1 = 8$ $499^1 = 499$ $(-3)^1 = -3$

Indices checklist

The base numbers have to be the same. ✓

If there's no index, the number has the power 1 ✓

Be careful with negatives: $(-3)^2 = 9$ ✓

Worked example

(a) Write $6 \times 6 \times 6 \times 6 \times 6$ as a single power of 6. **(1 mark)**

$6 \times 6 \times 6 \times 6 \times 6 = 6^5$

(b) Simplify $\dfrac{3^8 \times 3}{3^4}$ fully, leaving your answer in index form. **(2 marks)**

$\dfrac{3^8 \times 3}{3^4} = \dfrac{3^9}{3^4} = 3^5$

For 1(b), start by working out $\dfrac{9625}{7 \times 11}$

3 is the same as 3^1. For part (b), use the rule $a^m \times a^n = a^{m+n}$ to simplify the numerator, then use $\dfrac{a^m}{a^n} = a^{m-n}$ to simplify the fraction. Remember to write down both steps of your working and give your answer as a power.

Learn it!

You need to be able to recognise the square numbers up to 15^2 and the cubes of 1, 2, 3, 4, 5 and 10. You can check with a calculator but you'll be more confident if you learn them.

Now try this

1 (a) Write $7^3 \times 7^5$ as a single power of 7. **(1 mark)**

(b) $9625 = 5^n \times 7 \times 11$
Find the value of n. **(2 marks)**

2 $(\sqrt[3]{-27})^k = 9$
Write down the value of k. **(2 marks)**

3 (a) Simplify, leaving your answers in index form

(i) $\dfrac{2^9}{2^5}$ (ii) $(7^2)^6$ (iii) $5^2 \times 5^0$ **(3 marks)**

(b) $\dfrac{3^n}{3^2 \times 3^5} = 3^4$
Find the value of n. **(2 marks)**

Indices 2

You can use these index laws to deal with powers that are **fractions** or **negative numbers**.

① Negative powers

$$a^{-n} = \frac{1}{a^n}$$

$$5^{-2} = \frac{1}{5^2} = \frac{1}{25}$$

Be careful!

A **negative** power can still have a **positive** answer.

② Reciprocals

$$a^{-1} = \frac{1}{a}$$

This means that a^{-1} is the **reciprocal** of a.
You can find the reciprocal of a fraction by turning it upside down.

$$\left(\frac{5}{9}\right)^{-1} = \frac{9}{5}$$

③ Powers of fractions

$$\left(\frac{a}{b}\right)^n = \frac{a^n}{b^n}$$

$$\left(\frac{3}{10}\right)^2 = \frac{3^2}{10^2} = \frac{9}{100}$$

④ Combining rules

You can apply the rules one at a time.

$$\left(\frac{a}{b}\right)^{-n} = \left(\frac{b}{a}\right)^n = \frac{b^n}{a^n}$$

$$\left(\frac{2}{3}\right)^{-3} = \left(\frac{3}{2}\right)^3 = \frac{3^3}{2^3} = \frac{27}{8}$$

⑤ Fractional powers

You can use fractional powers to represent roots.

$$a^{\frac{1}{2}} = \sqrt{a} \qquad 49^{\frac{1}{2}} = 7$$
$$a^{\frac{1}{3}} = \sqrt[3]{a} \qquad 27^{\frac{1}{3}} = 3$$
$$a^{\frac{1}{4}} = \sqrt[4]{a} \qquad 16^{\frac{1}{4}} = 2$$

Check it!

A whole number raised to a power less than 1 gets smaller.

⑥ More complicated indices

You can use the index laws to work out more complicated fractional powers.

$$a^{\frac{m}{n}} = \left(\frac{1}{a^n}\right)^m$$

Do these calculations **one step at a time**.

$$27^{-\frac{2}{3}} = (27^{\frac{1}{3}})^{-2}$$
$$= (\sqrt[3]{27})^{-2}$$
$$= 3^{-2} = \frac{1}{3^2} = \frac{1}{9}$$

Worked example

Find the value of n when $3^n = 9^{-\frac{3}{2}}$
Show each step of your working clearly. **(3 marks)**

$$9^{-\frac{3}{2}} = (3^2)^{-\frac{3}{2}}$$
$$= 3^{2 \times -\frac{3}{2}} = 3^{-3}$$

So $3^n = 3^{-3}$ and $n = -3$

Problem solved!

3^n is not the same as $3n$. You can't divide by 3 to get n on its own. You need to make the base on the right-hand side the same as the base on the left-hand side.
1. Write 9 as a power of 3. Remember to use brackets.
2. Use $(a^n)^m = a^{nm}$ to write the right-hand side as a single power of 3.
3. Compare both sides and write down the value of n.

> You will need to use problem-solving skills throughout your exam – **be prepared!**

Now try this

1 You are given that $x = 7^h$ and $y = 7^k$
  Write each of the following as a single power of 7.
  (a) $\dfrac{x}{y}$ **(1 mark)**
  (b) x^2 **(1 mark)**
  (c) xy^2 **(2 marks)**

2 Given that $81^{-\frac{3}{4}} = 3^n$, find the value of n. **(3 marks)**

3 Write $\sqrt{\dfrac{49}{7^3}}$ as a single power of 7.
  Show every step of your working clearly. **(3 marks)**

Start by writing 49 as a power of 7

Calculator skills 1

These calculator keys are really useful.

| x^2 | Square a number. | $(-)$ | Enter a negative number. |

x^2 Square a number.

x^3 Cube a number.

x^{-1} Find the reciprocal of a number.

Ans   Use your previous answer in a calculation.

$(-)$ Enter a negative number.

$\sqrt{\square}$ Find the square root of a number.

$\sqrt[3]{\square}$ Find the cube root of a number. You might need to press the shift key first.

S⇔D   Change the answer from a fraction or surd to a decimal. Not all calculators have this key.

Rounding rules

1 To **round** a number, you look at the next digit on the right.

5 or more → round up   less than 5 → round down

2 Decimals can be rounded to a given number of **decimal places** (d.p.).

$6.475 = 6.48$ correct to 2 d.p.

3 To write a number correct to **3 significant figures** (3 s.f.), look at the fourth significant figure.

$0.003\,07\underline{9} = 0.003\,08$ to 3 s.f.

4 Leading zeros in decimals are not counted as significant.

5 Remember that the rule for significant figures still applies to **whole numbers**.

$27 = 30$ to 1 s.f.

Worked example

(a) Work out the value of $\dfrac{\sqrt{8.3}}{12.5 - 7.3}$

Give your answer as a decimal.
Write down all the figures on your calculator display. **(2 marks)**

$$\frac{\sqrt{8.3}}{12.5 - 7.3} = \frac{2.880\,97}{5.2} = 0.554\,033\,088$$

(b) Give your answer to part (a) correct to 2 significant figures. **(1 mark)**

0.55 (2 s.f.)

Calculate $\sqrt{8.3}$ using the $\sqrt{\square}$ key. Always show what the top of the fraction comes to as well as the bottom. Remember to write down **all** the figures on your calculator display.

Check it!

Do the whole calculation in one go on your calculator using the ▥ key.

Now try this

(a) Work out the value of $\dfrac{6.1 + 7.5}{1.8^2}$

Give your answer as a decimal.

Write down all the figures on your calculator display. **(2 marks)**

(b) Give your answer to part (a) correct to 3 significant figures. **(1 mark)**

Make sure you write down three significant figures even if the last digit is a zero.

Fractions

You need to be able to work with fractions and mixed numbers confidently **without a calculator**.

1 Adding or subtracting fractions

| Add or subtract the whole numbers |

$2\frac{2}{3} + 1\frac{1}{2}$

$= 3 + \frac{2}{3} + \frac{1}{2}$

| Write the fractions as fractions with the same denominator |

$= 3 + \frac{4}{6} + \frac{3}{6}$

$= 3 + \frac{7}{6}$

| Add or subtract the fractions |

$= 3 + 1\frac{1}{6}$

$= 4\frac{1}{6}$

| If you have an improper fraction then convert to a mixed number and add |

3 Dividing fractions

| Convert any mixed numbers to improper fractions |

$6\frac{1}{4} \div 1\frac{7}{8}$

$= \frac{25}{4} \times \frac{8}{15}$

| Turn the second fraction 'upside down' and change ÷ to × |

$= \frac{{}^{5}25 \times 8^{2}}{{}_{1}4 \times 15_{3}}$

$= \frac{10}{3}$

| Multiply the numerators and multiply the denominators, cancelling where possible |

$= 3\frac{1}{3}$

| Convert any improper fractions to mixed numbers |

2 Multiplying fractions

| Convert any mixed numbers to improper fractions |

$3\frac{1}{4} \times 2\frac{2}{3}$

$= \frac{13}{4} \times \frac{8}{3}$

| Multiply the numerators and multiply the denominators, cancelling where possible |

$= \frac{13 \times 8^{2}}{{}_{1}4 \times 3}$

$= \frac{26}{3}$

| Convert any improper fractions to mixed numbers |

$= 8\frac{2}{3}$

Worked example

Work out $7\frac{1}{3} - 2\frac{3}{4}$ **(3 marks)**

$7\frac{1}{3} - 2\frac{3}{4} = \frac{22}{3} - \frac{11}{4}$

$= \frac{88}{12} - \frac{33}{12}$

$= \frac{55}{12}$

$= 4\frac{7}{12}$

Remember you need to be able to do this **without** a calculator.

Worked example

The diagram shows three identical shapes.
$\frac{3}{4}$ of shape **A** is shaded and $\frac{3}{5}$ of shape **C** is shaded.

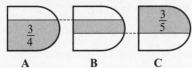

A   B   C

What fraction of shape **B** is shaded? **(3 marks)**

$1 - \frac{1}{4} - \frac{2}{5} = \frac{20}{20} - \frac{5}{20} - \frac{8}{20} = \frac{7}{20}$

Problem solved!

You need to **show your strategy**. This means you need to show all your working so it is clear how you have approached the problem.
White area on shape **A** $= 1 - \frac{3}{4} = \frac{1}{4}$
White area on shape **C** $= 1 - \frac{3}{5} = \frac{2}{5}$
So shaded area on shape **B** $= 1 - \frac{1}{4} - \frac{2}{5}$

| You will need to use problem-solving skills throughout your exam – **be prepared!** |

Now try this

1 Work out

(a) $\frac{7}{10} - \frac{1}{4}$ **(2 marks)**

(b) $3\frac{4}{9} + 1\frac{5}{6}$ **(3 marks)**

(c) $\frac{3}{4} \div \frac{5}{12}$ **(2 marks)**

(d) $1\frac{7}{8} \times 2\frac{2}{3}$ **(3 marks)**

2 Three girls shared a full bottle of cola.
Karen drank $\frac{1}{4}$ of the bottle.
Rita drank $\frac{3}{10}$ of the bottle.
Megan drank the rest.
(a) Work out the fraction of the bottle of cola which Megan drank. **(3 marks)**
Rita drank 36 cl of cola.
(b) How much cola was in the full bottle? **(2 marks)**

Worked solution video

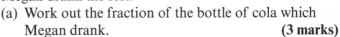

Decimals

Terminating decimals can be written exactly. You can write a terminating decimal as a fraction with denominator 10, 100, 1000, and so on.

$0.24 = \frac{24}{100} = \frac{6}{25}$

Recurring decimals have one digit or group of digits repeated forever. You can use dots to show the recurring digit or group of digits.

$\frac{2}{3} = 0.6666... = 0.\dot{6}$

> The dot tells you that the 6 repeats forever.

$\frac{346}{555} = 0.623\,4234... = 0.6\dot{2}3\dot{4}$

These dots tell you that the group of digits 234 repeats forever.

Recurring or terminating?

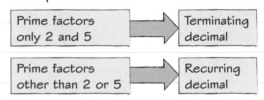

To check whether a fraction produces a recurring decimal or a terminating decimal, write it in its simplest form and find the prime factors of its **denominator**.

| Prime factors only 2 and 5 | → | Terminating decimal |
| Prime factors other than 2 or 5 | → | Recurring decimal |

> You could also write the denominator as a product of prime factors:
>
> $\frac{7}{50} = \frac{7}{2 \times 5^2}$
>
> The only factors are 2 and 5 so $\frac{7}{50}$ produces a terminating decimal.

Worked example

(a) Show that $\frac{7}{50}$ can be written as a terminating decimal. **(1 mark)**

$\frac{7}{50} = \frac{14}{100} = 0.14$

(b) Show that $\frac{11}{24}$ **cannot** be written as a terminating decimal. **(2 marks)**

$\frac{11}{24} = \frac{11}{2^3 \times 3}$

Denominator contains a factor other than 2 or 5 so decimal is recurring.

Worked example

(a) Show that $\frac{2}{9}$ is equivalent to 0.222... **(1 mark)**

$$\begin{array}{r} 0.2\ 2\ 2... \\ 9\overline{)2.^{2}0^{2}0^{2}0...} \end{array}$$

(b) Hence, or otherwise, write 0.7222... as a fraction. **(3 marks)**

$0.7222... = 0.222... + 0.5$
$= \frac{2}{9} + \frac{1}{2}$
$= \frac{4}{18} + \frac{9}{18} = \frac{13}{18}$

> You could also use long division for part (a).

Fractions and decimals

To convert a fraction into a decimal, divide the numerator by the denominator.

$\frac{2}{5} = 2 \div 5 = 0.4$

It's useful to remember these common fraction-to-decimal conversions:

Fraction	$\frac{1}{100}$	$\frac{1}{20}$	$\frac{1}{10}$	$\frac{1}{2}$	$\frac{1}{5}$	$\frac{1}{4}$	$\frac{3}{4}$
Decimal	0.01	0.05	0.1	0.5	0.2	0.25	0.75

Now try this

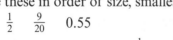

1 Write these in order of size, smallest first.

 0.6 $\frac{1}{2}$ $\frac{9}{20}$ 0.55   **(1 mark)**

2 Give evidence to show that $\frac{1}{250}$ can be written as a terminating decimal. **(2 marks)**

3 Show that $\frac{1}{140}$ **cannot** be written as a terminating decimal. **(2 marks)**

4 Write $\frac{5}{11}$ as a recurring decimal. **(2 marks)**

Estimation

You can estimate the answer to a calculation by rounding each number to **1 significant figure**, and then doing the calculation. You can use this method to check your answers, or to estimate calculations on your **non-calculator paper**. Here are two examples:

 $4.32 \times 18.09 \approx 4 \times 20 = 80$

The answer is approximately equal to 80.

 $327^2 \approx 300^2 = 3^2 \times 100^2 = 90\,000$

The answer is approximately equal to 90 000.

$\approx$ means 'is approximately equal to'

Decimal division trick

You might have to divide by a decimal on your non-calculator paper. If you multiply both numbers in a division by the same amount the answer stays the same.

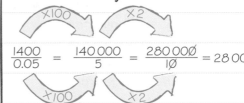

$$\frac{1400}{0.05} = \frac{140\,000}{5} = \frac{280\,000}{10} = 28\,000$$

Worked example

Work out an estimate for

(a) $\dfrac{4.31 \times 278}{0.487}$ **(2 marks)**

$$\frac{4.31 \times 278}{0.487} \approx \frac{4 \times 300}{0.5} = \frac{1200}{0.5} = 2400$$

(b) 37.4^3 **(2 marks)**

$$37.4^3 \approx 40^3 = 4^3 \times 10^3 = 64 \times 1000 = 64\,000$$

Round all the numbers to **1 significant figure**. Then **write out** the calculation with the rounded values before calculating your estimate.

You can use the laws of indices to work out 40^3 without a calculator.
$(ab)^n = a^n \times b^n$
so $40^3 = (4 \times 10)^3 = 4^3 \times 10^3$

Problem solved!

If you need to use formulae for spheres and cones they will be given with the question:

Surface area of sphere $= 4\pi r^2$

Volume of sphere $= \frac{4}{3}\pi r^3$

On your non-calculator paper you can use $\pi = 3.142$ then round to 1 s.f. to make your estimate.

You will need to use problem-solving skills throughout your exam – **be prepared!**

Worked example

A spherical ball-bearing has a radius of 2.35 cm.

(a) Work out an estimate for its surface area in square centimetres.   **(2 marks)**

$4\pi r^2 = 4 \times 3.142 \times 2.35^2$
$\approx 4 \times 3 \times 2^2 = 48 \text{ cm}^2$

(b) Is your answer to part (a) an overestimate or an underestimate? Give a reason for your answer.   **(1 mark)**

$3 < 3.142$ and $2 < 2.35$
so the answer is an underestimate.

Now try this

1 Showing your rounding, work out an estimate for

$$\frac{82 \times 285}{64 \times 35}$$ **(2 marks)**

2 A scientist models a raindrop as a sphere with radius 3.2 mm.
Volume of a sphere $= \frac{4}{3}\pi r^3$

(a) Work out an estimate for the volume of the raindrop.   **(2 marks)**

(b) Is your answer to part (a) an overestimate or an underestimate?
Give a reason for your answer.   **(1 mark)**

Standard form

Numbers in standard form have two parts.

$$7.3 \times 10^{-6}$$

This part is a number greater than or equal to 1 and less than 10

This part is a power of 10

You can use standard form to write very large or very small numbers.

$$920\,000 = 9.2 \times 10^5$$

Numbers greater than 10 have a positive power of 10

$$0.007\,03 = 7.03 \times 10^{-3}$$

Numbers less than 1 have a negative power of 10

Counting decimal places

You can count decimal places to convert between numbers in standard form and ordinary numbers.

3 jumps

$$7\,9\,0\,0 = 7.9 \times 10^3$$

7900 > 10
So the power is positive

4 jumps

$$0.0\,0\,0\,3\,5 = 3.5 \times 10^{-4}$$

0.00035 < 1
So the power is negative

Be careful!
Don't just count zeros to work out the power.

Worked example

Work out the value of $(8.3 \times 10^6) - (4.1 \times 10^5)$
Give your answer in standard form.   **(2 marks)**

$$\begin{array}{r} {}^{7\ \ 12\ \ 1}\\ \cancel{8}\,300\,000 \\ -\ \ \ 410\,000 \\ \hline 7\,890\,000 = 7.89 \times 10^6 \end{array}$$

This is a **non-calculator** question. To add or subtract numbers in standard form, write them as ordinary numbers first. Then use a written method to add or subtract.

Non-calculator multiplying

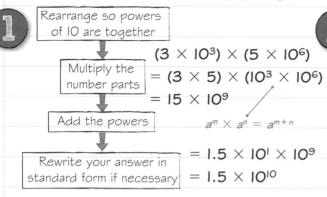

1
Rearrange so powers of 10 are together

Multiply the number parts

Add the powers

Rewrite your answer in standard form if necessary

$(3 \times 10^3) \times (5 \times 10^6)$
$= (3 \times 5) \times (10^3 \times 10^6)$
$= 15 \times 10^9$

$a^m \times a^n = a^{m+n}$

$= 1.5 \times 10^1 \times 10^9$
$= 1.5 \times 10^{10}$

Non-calculator dividing

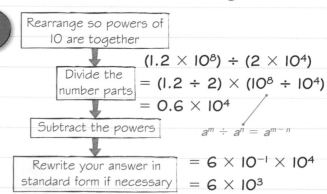

2
Rearrange so powers of 10 are together

Divide the number parts

Subtract the powers

Rewrite your answer in standard form if necessary

$(1.2 \times 10^8) \div (2 \times 10^4)$
$= (1.2 \div 2) \times (10^8 \div 10^4)$
$= 0.6 \times 10^4$

$a^m \div a^n = a^{m-n}$

$= 6 \times 10^{-1} \times 10^4$
$= 6 \times 10^3$

Using a calculator

You can enter numbers in standard form using the ×10ˣ key.

To enter 3.7×10^{-6} press

[3] [.] [7] [×10ˣ] [(−)] [6]

If you are using a calculator with numbers in standard form it is a good idea to put brackets around each number.

Now try this

The mass of one E.coli bacteria is 6×10^{-16} grams. Find the total mass of 3×10^6 bacteria?   **(2 marks)**

Have a go at this question **without a calculator** first. Then use your calculator to check your answer.

Recurring decimals

You can use algebra to convert a recurring decimal into a fraction. Here is the strategy:

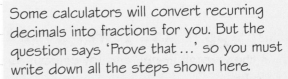

Write the recurring decimal as n. → Multiply by 10, 100 or 1000. → Subtract to remove the recurring part. → Divide by 9, 99 or 999 to write as a fraction.

If you need to do this in your exam you must show **all your working**. For a reminder about recurring decimals have a look at page 6.

Worked example

Prove that the recurring decimal $0.\dot{2}\dot{4}$ has the value $\frac{8}{33}$ **(2 marks)**

Let $n = 0.242\,424\,24\ldots$

$100n = 24.242\,424\,24\ldots$

$-\ n = 0.242\,424\,24\ldots$

$99n = 24$

$n = \dfrac{24}{99} = \dfrac{8}{33}$

Some calculators will convert recurring decimals into fractions for you. But the question says 'Prove that ...' so you must write down all the steps shown here.

1. Write the recurring decimal equal to n, and write out some of its digits.
2. Multiply both sides by 100 as there are 2 recurring digits.
3. Subtract n to remove the recurring part.
4. Divide both sides by 99 to write n as a fraction.
5. Simplify the fraction.

Multiply by...

10 if 1 digit recurs.

100 if 2 digits recur.

1000 if 3 digits recur.

Worked example

Show that $0.4\dot{7}\dot{3}$ can be written as the fraction $\frac{469}{990}$ **(2 marks)**

Let $n = 0.473\,737\,37\ldots$

$100n = 47.373\,737\,37\ldots$

$-\ n = 0.473\,737\,37\ldots$

$99n = 46.9$

$n = \dfrac{46.9}{99}$

$n = \dfrac{469}{990}$

Problem solved!

In this recurring decimal the digit 4 does not recur. Follow the same steps to write n as a fraction. After you divide by 99, multiply the top and bottom of your fraction by 10 to convert the decimal in the numerator into an integer.

You will need to use problem-solving skills throughout your exam – **be prepared!**

Worked solution video

$0.\dot{3}5\dot{1} = 0.351351351\ldots$

There are 3 recurring digits so you need to write $0.\dot{3}5\dot{1}$ as n, then multiply by 1000. You will get a fraction with denominator 999 which you can simplify.

Now try this

1 Work out the recurring decimal $0.\dot{5}\dot{4}$ as a fraction in its simplest form. **(2 marks)**

2 Prove that the recurring decimal $0.01\dot{8}$ has the value $\frac{1}{55}$ **(2 marks)**

3 Show that $0.\dot{3}5\dot{1}$ can be written as the fraction $\frac{13}{37}$ **(2 marks)**

Upper and lower bounds

Upper and lower bounds are a measure of accuracy. For example, the width of a postcard is given as 8 cm to the nearest cm.

```
     lower            upper
     bound            bound
  +────┼────┼────┼────┼────+
 7 cm 7.5 cm 8 cm 8.5 cm 9 cm
       ←─────────→
```

The actual width of the postcard could be anything between 7.5 cm and 8.5 cm.

7.5 cm is called the **lower bound**.

8.5 cm is called the **upper bound**.

Using upper and lower bounds in calculations

To find the overall upper and lower bounds of the answer to a calculation use these rules.

	+	−	×	÷
Overall upper bound	UB + UB	UB − LB	UB × UB	UB ÷ LB
Overall lower bound	LB + LB	LB − UB	LB × LB	LB ÷ UB

Overall lower bound of a + b = lower bound of a + lower bound of b

Worked example

A roll of ribbon is 100 cm long, correct to 2 significant figures.
A 21-cm piece of ribbon is cut off the roll, correct to the nearest cm.
Calculate the lower bound, in cm, for the amount of ribbon remaining on the roll.   **(3 marks)**

	Lower bound	Upper bound
Length of ribbon	95 cm	105 cm
Length of piece cut off	20.5 cm	21.5 cm

Lower bound of remaining length
= 95 − 21.5
= 73.5 cm

If you're answering questions about upper and lower bounds, it's a good idea to write out the upper bound and lower bound for **all values** given in the question before you start. To work out the **lower bound** for a − b you need to use the **lower bound** for a and the **upper bound** for b.

Different values might be given to **different degrees of accuracy**. The length of the roll is correct to 2 significant figures, or to the nearest **10 cm**. The length of the piece cut off is correct to the nearest **cm**. Be really careful when you're working out your upper and lower bounds.

Now try this

1   The area of a rectangle is 320 cm². The length of the rectangle is 22 cm. Both values are correct to 2 significant figures. Calculate the lower bound for the width of the rectangle. Show your working clearly.   **(3 marks)**

2   Correct to 2 decimal places, the volume of a solid cube is 3.37 m³. Calculate the upper bound for the surface area of the cube.   **(4 marks)**

For a cube with edges of length x, the volume is x³ and the surface area is 6x²

Accuracy and error

When a question involves **upper and lower bounds**, you might need to give your answer to an **appropriate** degree of accuracy. You can use this strategy:

Calculate the **lower bound** for the answer.	→	Calculate the **upper bound** for the answer.	→	Choose the most accurate value that **both bounds** would round to.

If you're not confident with upper and lower bounds, revise them now on page 10.

Worked example

A cylinder has a volume of $115\,\text{cm}^3$, to the nearest cm^3. Its radius is $2.3\,\text{cm}$, correct to 1 decimal place. Find the height of the cylinder to an appropriate degree of accuracy.
You must explain why your answer is to an appropriate degree of accuracy. **(4 marks)**

	Upper bound	Lower bound
Volume	$115.5\,\text{cm}^3$	$114.5\,\text{cm}^3$
Radius	$2.35\,\text{cm}$	$2.25\,\text{cm}$

UB for height $= \dfrac{115.5}{\pi \times 2.25^2} = 7.26218\ldots$

LB for height $= \dfrac{114.5}{\pi \times 2.35^2} = 6.59963\ldots$

Height $= 7\,\text{cm}$ (1 s.f.)

Problem solved!

If a question asks for an **appropriate degree of accuracy** you will have to consider upper and lower bounds. You need to choose an answer that **both** bounds will round to.
Check 2 significant figures:
UB = 7.3 cm (2 s.f.) and LB = 6.6 cm (2 s.f.) ✗
Check 1 significant figure:
UB = 7 cm (1 s.f.) and LB = 7 cm (1 s.f.) ✓
Lay out your working clearly so that you can show your method. You could also write a sentence like this at the end of your answer:
'UB and LB both round to 7 cm to 1 s.f.'

You will need to use problem-solving skills throughout your exam – **be prepared!**

Writing error intervals

You can streamline your answers by using **inequality notation** to write down upper and lower bounds.

$x = 4.7$ (2 s.f.) means $4.65 \leqslant x < 4.75$
$y = 800$ (1 s.f.) means $750 \leqslant y < 850$
For the worked example above you could write:

$7.26218\ldots \leqslant$ Height $< 6.59963\ldots$

Use 'less than or equal to' for the lower bound.   Use 'less than' for the upper bound.

You will need to use the formula for the area of a triangle that is on page 101. You want to know whether the farmer definitely has enough bags, so consider the **worst-case scenario**: the UB for the area of the field and the LB for the size of one bag.

You can revise distance, speed and time on page 67.

Now try this

1 A go-karting course is 425 m long, correct to the nearest metre.
   Nisha completes the course in 1 minute 25 seconds, to the nearest second.
   Calculate her average speed in m/s to an appropriate degree of accuracy.
   You must explain why your answer is to an appropriate degree of accuracy. **(4 marks)**

2 The diagram shows a farmer's field.

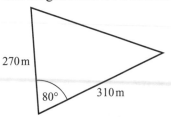

Worked solution video

The lengths of the sides of the field have been measured to the nearest 10 m. The farmer wants to plant grass in the field.
1 bag of seed covers $500\,\text{m}^2$, to the nearest $100\,\text{m}^2$.
The farmer has 90 bags of grass seed.
Does the farmer definitely have enough bags?
Show all your working. **(4 marks)**

Surds 1

You can give exact answers to calculations by leaving some numbers as square roots.

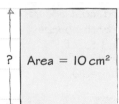

? | Area = 10 cm²

This square has a side length of $\sqrt{10}$ cm.

You can't write $\sqrt{10}$ exactly as a decimal number. It is called a **surd**.

Rules for simplifying square roots

These are the most important rules to remember when dealing with surds.

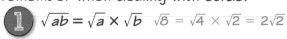

1 $\sqrt{ab} = \sqrt{a} \times \sqrt{b}$ $\sqrt{8} = \sqrt{4} \times \sqrt{2} = 2\sqrt{2}$

2 $\sqrt{\dfrac{a}{b}} = \dfrac{\sqrt{a}}{\sqrt{b}}$ $\sqrt{\dfrac{3}{25}} = \dfrac{\sqrt{3}}{\sqrt{25}} = \dfrac{\sqrt{3}}{5}$

You need to remember these rules for your exam.

Worked example

Show that $\sqrt{45} = 3\sqrt{5}$
Show each stage of your working clearly.   **(2 marks)**

$\sqrt{45} = \sqrt{9 \times 5}$
$= \sqrt{9} \times \sqrt{5}$
$= 3\sqrt{5}$

This question says 'Show that …' so you can't use your calculator. You need to show each step of your working clearly:

1. Look for a factor of 45 which is a square number: $45 = 9 \times 5$
2. Use the rule $\sqrt{ab} = \sqrt{a} \times \sqrt{b}$ to split the square root into two square roots.
3. Write $\sqrt{9}$ as a whole number.

Rationalising the denominator of a fraction means making the denominator a whole number.

You can do this by multiplying the top **and** bottom of the fraction by the surd part in the denominator.

$\times\sqrt{2}$

$\dfrac{5}{3\sqrt{2}} = \dfrac{5\sqrt{2}}{6}$

$\times\sqrt{2}$

The surd part of the denominator is $\sqrt{2}$

Remember that $\sqrt{2} \times \sqrt{2} = 2$
So $3\sqrt{2} \times \sqrt{2} = 3 \times 2 = 6$

Good form

Most surd questions ask you to write a number or answer in a certain **form**.

This means you need to find **integers** for all the letters in the expression.

$6\sqrt{3}$ is in the form $k\sqrt{3}$
$k = 6$

The integers can be positive or negative.

$4 - 9\sqrt{2}$ is in the form $p + q\sqrt{2}$

$p = 4$ and $q = -9$

You can check your answer by writing down the integer value for each letter.

Now try this

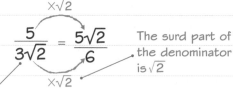

Find factors of 32 and 98 which are **square** numbers.

1 Write $\sqrt{32} + \sqrt{98}$ in the form $p\sqrt{2}$ where p is an integer. Show each stage of your working clearly.   **(2 marks)**

2 Show that $\dfrac{35}{\sqrt{7}} = 5\sqrt{7}$ **(2 marks)**

Rationalise the denominator by multiplying top and bottom by $\sqrt{7}$

3 x is an integer such that
$$\dfrac{\sqrt{x} \times \sqrt{18}}{\sqrt{3}} = 8\sqrt{3}$$
Find the value of x.   **(4 marks)**

Counting strategies

You might need to find strategies for counting the total number of possible **combinations**. One way of finding combinations is to make a systematic list. Here are all the possible three-digit numbers that can be made from the number cards shown on the right.

$\boxed{4}$ $\boxed{5}$ $\boxed{6}$

456 465   **546 564**   **645 654** •————————•There are six possibilities

Start by writing out the numbers that begin with 4.

Then write the numbers beginning with 5.

Finally, write the numbers that begin with 6.

Counting with calculation

Sometimes it's impossible to write down every possible combination. You can count combinations by **multiplying** the number of choices for each option. Here is a four-character password for a website:

First character must be a letter: 26 choices

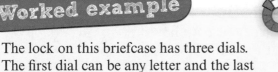

Fourth character must be one of: $, ·, _ or # 4 choices

Second character must be a digit from 0 to 9: 10 choices

Third character must be a letter: 26 choices

The total number of possible passwords is 26 × 10 × 26 × 4 = 27 040.

The lock on this briefcase has three dials. The first dial can be any letter and the last two dials can be any digit from 0 to 9. Here is one possible combination. $\boxed{M}\boxed{7}\boxed{0}$

(a) How many different ways are there of setting the code?   **(2 marks)**

26 × 10 × 10 = 2600

A different suitcase has four dials. The first two dials can be any letter from A to E, and the last two dials can be any **even** digit greater than 0. Here is one possible combination. $\boxed{E}\boxed{B}\boxed{4}\boxed{2}$

(b) How many different ways are there of setting this code?   **(2 marks)**

5 × 5 × 4 × 4 = 400

Work out how many choices there are for each dial.

Dial 1: Letter = 26 choices

Dial 2: Digit from 0 to 9 = 10 choices

Dial 3: Digit from 0 to 9 = 10 choices

Multiply the number of choices to work out the total number of possible combinations.

Problem solved!

The number of choices for each dial has **changed** here, so be really careful.

Dials 1 and 2: Letter from A to E = 5 choices

Dials 3 and 4: Digits 2, 4, 6 or 8 = 4 choices

It's important to **show your strategy** as well as your answer, so write out the calculation you use.

You will need to use problem-solving skills throughout your exam – **be prepared!**

To play the Euromillions lottery you select 5 numbers from 1 to 50. You also select 2 lucky stars from 1 to 11.

Work out how many different ways there are of selecting these 7 numbers. Give your answer in standard form, correct to 3 significant figures.   **(2 marks)**

Problem-solving practice 1

Throughout your Higher GCSE exam you will need to **problem-solve**, **reason**, **interpret** and **communicate** mathematically. If you come across a tricky or unfamiliar question in your exam you can try some of these strategies:

☑ Sketch a diagram to see what is going on.

☑ Try the problem with smaller or easier numbers.

☑ Plan your strategy before you start.

☑ Write down any formulae you might be able to use.

☑ Use x or n to represent an unknown value.

---

1 The diagram shows two types of plastic building block.

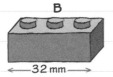

A   B

←24 mm→   ←—32 mm—→

Worked solution video

Block **A** is 24 mm long.

Block **B** is 32 mm long.

Jeremy joins some type **A** blocks together to make a straight row.

He then joins some type **B** blocks together to make a straight row of the same length.

Write down the shortest possible length of this row. **(4 marks)**

Factors and primes page 1

You have to use a whole number of building blocks in each row, so the length of each row has to be a multiple of the length of one block. The answer will be the lowest common multiple of 24 and 32. You can't get all the marks just by writing down the answer. You need to show how you found the answer clearly and neatly.

TOP TIP

If you're not sure how to start, draw a sketch. This might help you see that the lengths are multiples of 24 and 32.

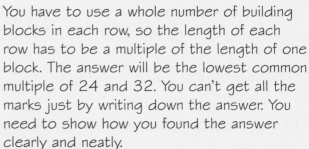

---

2 Susan has 2 dogs.

Each dog is fed $\frac{3}{8}$ kg of dog food each day.

Susan buys dog food in bags.

Each bag weighs 14 kg.

For how many days can Susan feed the 2 dogs from 1 bag of dog food?

You must show **all** your working. **(5 marks)**

Fractions page 5

There are lots of steps in this question so make sure you keep track of your working and write it down clearly.

TOP TIP

Write words with each calculation to explain what you are doing.

---

3 Prove that the recurring decimal $0.9\dot{2}\dot{8}$ can be written as $\frac{919}{990}$

(3 marks)

Worked solution video

Recurring decimals page 9

Start by writing $x = 0.9282828...$ If you work out $1000x$ and $10x$ you can keep all your working in whole numbers.

TOP TIP

If a question says 'Show that …' or 'Prove that …' you have to show every step of your working clearly and neatly.

Problem-solving practice 2

4   (a) $a = 4 \times 10^{2n}$ where n is an integer.

Find, in standard form, an expression for $\sqrt{a}$ **(2 marks)**

(b) $b = 8 \times 10^{3m}$ where m is an integer.

Find, in standard form, an expression for $b^{\frac{4}{3}}$ **(3 marks)**

Standard form page 8
Indices 2 page 3

Remember that $(xy)^n = x^n y^n$

Be careful in part (b): the first part of a number in standard form must be greater than or equal to 1 and less than 10.

TOP TIP

Questions on indices can involve unknowns, so your calculator won't be able to help you. Make sure you know the laws of indices.

---

5   The diagram shows a wooden planting box in the shape of a cuboid.

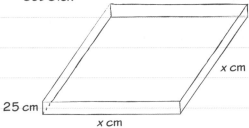

The volume of the box is $810\,000\,cm^3$ correct to 2 significant figures.

The depth of the box is 25 cm, to the nearest cm.

The box has a square base with sides of length x cm.

Find the lower bound for x. Give your answer correct to 3 significant figures.

(4 marks)

Upper and lower bounds page 10
Volumes of 3D shapes page 85

Complete this table showing the upper and lower bounds for each measurement before you start:

	25 cm	810 000 cm
Upper	25.5 cm	
Lower		805 000 cm

You are **dividing** the volume by the depth to work out x^2. Choose the values you use carefully to make the answer as **small** as possible.

TOP TIP

When answering questions about upper and lower bounds, it's a good idea to write out the upper and lower bounds for all the values before you start.

---

6   (a) How many numbers are there between 1000 and 9999 that contain only odd digits?   **(2 marks)**

(b) In a popular word game, there are 1310 allowable 3-letter words.
A monkey types 3 letters at random. Calculate the probability that it types a word that is allowable in the game.   **(3 marks)**

Counting strategies page 13
Probability page 123

Think about the number of choices you have for each digit, or for each letter. You multiply the number of choices for each option to work out the total possible number of combinations.

TOP TIP

Remember to write down any calculations you carry out so you can show your strategy.

Algebraic expressions

You need to be able to work with algebraic expressions confidently. For a reminder about using the index laws with **numbers** have a look at pages 2 and 3.

 1 You can use the **index laws** to simplify algebraic expressions.

$a^m \times a^n = a^{m+n}$

$x^4 \times x^3 = x^{4+3} = x^7$

$\dfrac{a^m}{a^n} = a^{m-n}$

$m^8 \div m^2 = m^{8-2} = m^6$

$(a^m)^n = a^{mn}$

$(n^2)^4 = n^{2\times4} = n^8$

2 You can square or cube a whole expression.

$(4x^3y)^2 = (4)^2 \times (x^3)^2 \times (y)^2$
$= 16x^6y^2$

$16 = (4)^2$

You need to square everything inside the brackets.

$(x^3)^2 = x^{3\times2} = x^6$

Remember that if a letter appears on its own then it has the power 1.

3 Algebraic expressions may also contain negative and fractional indices.

$a^{-m} = \dfrac{1}{a^m}$

$(c^2)^{-3} = c^{2\times-3} = c^{-6} = \dfrac{1}{c^6}$

$a^{\frac{1}{n}} = \sqrt[n]{a}$

$(8p^3)^{\frac{1}{3}} = (8)^{\frac{1}{3}} \times (p^3)^{\frac{1}{3}}$
$= \sqrt[3]{8} \times p^{3\times\frac{1}{3}}$
$= 2p$

One at a time

When you are **multiplying** expressions:

1. Multiply any number parts first.
2. Add the powers of each letter to work out the new power.

$6p^2q \times 3p^3q^2 = 18p^5q^3$

$6 \times 3 = 18$

$p^2 \times p^3 = p^{2+3} = p^5$

$q \times q^2 = q^{1+2} = q^3$

When you are **dividing** expressions:

1. Divide any number parts first.
2. Subtract the powers of each letter to work out the new power.

$12 \div 3 = 4$

$b^3 \div b^2 = b^{3-2} = b$

$\dfrac{12a^5b^3}{3a^2b^2} = 4a^3b$

$a^5 \div a^2 = a^{5-2} = a^3$

Worked example

Simplify fully

(a) $m \times m \times m \times m$ **(1 mark)**

m^4

(b) $(x^3)^3$ **(1 mark)**

x^9

(c) $\dfrac{4y^2 \times 3y^7}{6y}$ **(2 marks)**

$\dfrac{4y^2 \times 3y^7}{6y} = \dfrac{12y^9}{6y} = 2y^8$

(a) $m = m^1$, so $m \times m \times m \times m$
$= m^{1+1+1+1}$

(b) Use $(a^m)^n = a^{mn}$

(c) Start by simplifying the top part of the fraction. Do the number part first then the powers. Use $a^m \times a^n = a^{m+n}$

Next divide the expressions. Divide the number part, then divide the indices using $\dfrac{a^m}{a^n} = a^{m-n}$

Worked solution video

Now try this

 1 Simplify $(h^2)^6$ **(1 mark)**

2 Simplify fully

(a) $(2a^5 b)^4$ **(2 marks)**

(b) $5x^4y^2 \times 3x^3y^7$ **(2 marks)**

(c) $18d^8g^{10} \div 6d^2g^5$ **(2 marks)**

 3 (a) Simplify $(16p^{10})^{\frac{1}{2}}$ **(2 marks)**

(b) Simplify $(64x^9y^2)^{-\frac{1}{3}}$ **(2 marks)**

 Apply the power outside the brackets to everything inside the brackets.

Expanding brackets

Expanding or multiplying out brackets is a key algebra skill.

You have to multiply the expression outside the bracket by everything inside the bracket.

$4n \times n^2 = 4n^3$

$$4n(n^2 + 2) = 4n^3 + 8n$$

$4n \times 2 = 8n$

'Expand and simplify' means 'multiply out and then collect like terms'.

Golden rule

When you expand, you need to be careful with negative signs in front of the bracket.

Negative signs belong to the term to their right.

$-2 \times x \qquad -2 \times -y$

$$x - 2(x - y) = x - 2x + 2y$$
$$= -x + 2y$$

Multiply out the brackets first and then collect like terms if possible.

You can use the **grid method** to expand two brackets.

$(x + 7)(x - 5) = x^2 - 5x + 7x - 35$
$$= x^2 + 2x - 35$$

Remember to collect like terms if possible.

	x	-5
x	x^2	$-5x$
7	$7x$	-35

The negative sign belongs to the 5.
You need to write it in your grid.

Or

You can use the acronym FOIL to expand two brackets.

$2a^2 \qquad -b^2$

$$(2a + b)(a - b) = 2a^2 - 2ab + ab - b^2$$
$$= 2a^2 - ab - b^2$$

ab

$-2ab$

First terms
Outer terms
Inner terms
Last terms

Some people remember this as a 'smiley face'.

Worked example

(a) Expand and simplify $(3p - 4)^2$ **(2 marks)**

$(3p - 4)^2 = (3p - 4)(3p - 4)$
$$= 9p^2 - 12p - 12p + 16$$
$$= 9p^2 - 24p + 16$$

	$3p$	-4
$3p$	$9p^2$	$-12p$
-4	$-12p$	16

Be careful with the negative signs:
$-4 \times -4 = 16 \qquad p \times -4 = -4p$

You might have to multiply three factors together. Start by expanding $(x - 2)(x + 3)$ and remember to write brackets around the whole expansion. Then multiply the term outside the brackets, x, by **every term** inside the brackets.

(b) Expand and simplify $x(x - 2)(x + 3)$ **(2 marks)**

$x(x - 2)(x + 3) = x(x^2 + 3x - 2x - 6)$
$$= x(x^2 + x - 6)$$
$$= x^3 + x^2 - 6x$$

Check it!

Try an easy value, like $x = 5$
$x(x - 2)(x + 3) = 5 \times 3 \times 8 = 120$
$x^3 + x^2 - 6x = 125 + 25 - 30 = 120$ ✓

Now try this

Worked solution video

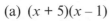

1 Expand and simplify
   (a) $(x + 5)(x - 1)$ **(2 marks)**
   (b) $(p - 6)^2$ **(2 marks)**

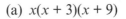

2 Expand and simplify
   (a) $x(x + 3)(x + 9)$ **(2 marks)**
   (b) $(n + 5)(n + 3)^2$ **(3 marks)**

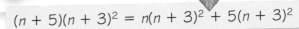
$(n + 5)(n + 3)^2 = n(n + 3)^2 + 5(n + 3)^2$

Factorising

Factorising is the opposite of expanding brackets:

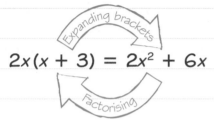

$$2x(x + 3) = 2x^2 + 6x$$

You need to look for the **largest factor** you can take out of every term in the expression.

$$10a^2 + 5ab = 5(2a^2 + ab)$$

This expression has only been **partly factorised**.

$$10a^2 + 5ab = 5a(2a + b)$$

This expression has been **completely factorised**.

Factorising $x^2 + bx + c$

You need to write the expression with **two brackets**.

You need to find two numbers which add up to 7... $5 + 2 = 7$

$$x^2 + 7x + 10 = (x + 5)(x + 2)$$

... and multiply to make 10 $5 \times 2 = 10$

When factorising $x^2 + bx + c$, use this table to help you find the two numbers:

b	c	Factors
Positive	Positive	Both numbers positive
Positive	Negative	Bigger number positive and smaller number negative
Negative	Negative	Bigger number negative and smaller number positive
Negative	Positive	Both numbers negative

Factorising $ax^2 + bx + c$

$$2x^2 - 7x - 15 = (2x \quad)(x \quad)$$

One of the brackets must contain a $2x$ term. Try pairs of numbers which have a product of -15. Check each pair by multiplying out the brackets.

$(2x + 5)(x - 3) = 2x^2 - x - 15$ ✗

$(2x - 3)(x + 5) = 2x^2 + 7x - 15$ ✗

$(2x + 3)(x - 5) = 2x^2 - 7x - 15$ ✓

Difference of two squares

You can factorise expressions that are written as

$$(\text{something})^2 - (\text{something else})^2$$

Use this rule:

$$a^2 - b^2 = (a + b)(a - b)$$

$$x^2 - 36 = x^2 - (6)^2$$
$$= (x + 6)(x - 6)$$

36 is a square number.

$36 = 6^2$ so $a = x$ and $b = 6$

Worked example

 Factorise fully

(a) $p^2 - 3p$ **(2 marks)**

 (b) $15x^2 + 5xy$ **(2 marks)**

$p(p - 3)$

$5x(3x + y)$

You need to look for the **largest** factor you can take out of every term.

Partly factorised: $x(15x + 5y)$ ✗

Partly factorised: $5(3x^2 + xy)$ ✗

Fully factorised: $5x(3x + y)$ ✓

Now try this

1 Factorise

  (a) $4a - 6$ **(1 mark)**

  (b) $y^2 + 5y$ **(1 mark)**

2 Factorise fully

  (a) $12g + 3g^2$ **(2 marks)**

  (b) $p^2 - 15p + 14$ **(2 marks)**

  (c) $6x^2 - 8xy$ **(2 marks)**

3 Factorise

  (a) $4ma - 24m^2a$ **(2 marks)**

  (b) $p^2 - 64$ **(1 mark)**

4 Factorise $3x^2 - 8x + 4$ **(2 marks)**

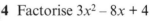

Worked solution video

Linear equations 1

To solve a linear equation you need to get the letter on its own on one side.
It is really important to write your working **neatly** when you are solving equations.

Every line of working should have an equals sign in it.

Start a new line for each step.
Do one operation at a time.

$$5x + 3 = 18 \quad (-3)$$
$$5x = 15 \quad (\div 5)$$
$$x = 3$$

Write down the operation you are carrying out. Remember to do the same thing to both sides of the equation.

Line up the equals signs.

Letter on both sides?

To solve an equation you have to get the letter on its own on one side of the equation.

Start by collecting like terms so that all the letters are together.

$$2 - 2x = 26 + 4x \quad (+ 2x)$$
$$2 = 26 + 6x \quad (- 26)$$
$$-24 = 6x \quad (\div 6)$$
$$-4 = x$$

You can write your answer as

$-4 = x$ or as $x = -4$

Equations with brackets

Always start by multiplying out the brackets then collecting like terms.

For a reminder about multiplying out brackets have a look at page 17.

$$19 = 8 - 2(5 - 3y)$$
$$19 = 8 - 10 + 6y$$
$$19 = -2 + 6y \quad (+ 2)$$
$$21 = 6y \quad (\div 6)$$
$$\frac{21}{6} = y$$
$$y = \frac{7}{2} \text{ or } 3\frac{1}{2} \text{ or } 3.5$$

Your answer can be written as a fraction or decimal.

Worked example

Solve $7r + 2 = 5(r - 4)$ **(3 marks)**

$$7r + 2 = 5r - 20 \quad (- 5r)$$
$$2r + 2 = -20 \quad (- 2)$$
$$2r = -22 \quad (\div 2)$$
$$r = -11$$

Multiply out the brackets then collect all the terms in r on one side. You need to write down each step of your working clearly.

Check it!

Substitute $r = -11$ into each side of the equation.

Left-hand side: $7(-11) + 2 = -75$

Right-hand side: $5(-11 - 4) = -75$ ✓

Now try this

1 Solve
  (a) $5w - 17 = 2w + 4$ **(3 marks)**
  (b) $2(x + 11) = 20$ **(3 marks)**

2 Solve
  (a) $6y - 9 = 2(y - 8)$ **(3 marks)**
  (b) $4m - 2(m - 3) = 7m - 14$ **(3 marks)**

Expand the brackets first.

Expand the brackets then collect all the m terms on one side of the equation.

Linear equations 2

Equations with fractions

When you have an equation with fractions, you need to get rid of any fractions before solving. You can do this by multiplying every term by the lowest common multiple (LCM) of the denominators.

The LCM of 3 and 5 is 15.

$$\frac{x}{3} + \frac{x-1}{5} = 11 \qquad (\times 15)$$

$$\frac{^5\cancel{15}x}{\cancel{3}_1} + \frac{^3\cancel{15}(x-1)}{\cancel{5}_1} = 165$$

Cancel the fractions. There is more about simplifying algebraic fractions on page 48.

$$5x + 3x - 3 = 165$$
$$8x - 3 = 165 \qquad (+3)$$
$$8x = 168 \qquad (\div 8)$$
$$x = 21$$

Multiplying by an expression

You might have to multiply by an expression to get rid of the fractions.

$$\frac{20}{n-3} = -5 \qquad (\times(n-3))$$
$$20 = -5(n-3)$$

Worked example

Solve $\dfrac{29-x}{4} = x + 5$ **(3 marks)**

$$\frac{4(29-x)}{4} = 4(x+5)$$
$$29 - x = 4(x+5)$$
$$29 - x = 4x + 20 \qquad (+x)$$
$$29 = 5x + 20 \qquad (-20)$$
$$9 = 5x \qquad (\div 5)$$
$$\frac{9}{5} = x$$

Eliminate fractions **before** you start solving the equation. You can do this by multiplying both sides of the equation by 4.
Use brackets to show that you are multiplying everything by 4.
$4(x + 5)$ ✓ $4x + 5$ ✗
Multiply out the brackets, then solve the equation normally. Remember that your answer could be a fraction.

Top tip!

It's OK to leave the answer to an equation as an improper fraction. Don't waste time converting to mixed numbers or decimals.

Writing your own equations

You can find unknown values by writing and solving equations.

$4(x - 1)$ cm $(3x + 3)$ cm

$\frac{5}{n}$ m

Perimeter = 20 m   2 m

$$4(x - 1) = 3x + 3 \qquad \frac{5}{n} + \frac{5}{n} + 2 + 2 = 20$$

Now try this

1 Solve

(a) $\dfrac{25 - 3w}{4} = 10$ **(3 marks)**

(b) $5x - 10 = \dfrac{18 - x}{3}$ **(3 marks)**

2 Solve

(a) $\dfrac{2y}{3} + \dfrac{y - 4}{2} = 5$ **(3 marks)**

(b) $\dfrac{3m - 1}{4} - \dfrac{2m + 4}{3} = 1.5$ **(3 marks)**

Formulae

A **formula** is a mathematical rule.

You can write formulae using algebra.

This label shows a formula for working out the cooking time of a chicken.

You can write this formula using algebra as

FREE-RANGE CHICKEN		
WEIGHT (KG)	**PRICE PER KG**	**COOKING INSTRUCTIONS**
1.8	£3.95	Cook at 170°C for 25 minutes per kg plus half an hour

$T = 25w + 30$, where T is the cooking time in minutes and w is the weight in kg.

In the description of each variable, you must give the units.

If T was the cooking time in hours then this formula would give you a very crispy chicken!

Worked example

This formula is used in physics to calculate distance:

$D = ut - 5t^2$

$u = 14$ and $t = -3$

Work out the value of D.   **(2 marks)**

$D = (14)(-3) - 5(-3)^2$

$\quad = (14)(-3) - 5(9)$

$\quad = -42 - 45$

$\quad = -87$

Substitute the values for u and t into the formula.

If you use brackets then you're less likely to make a mistake. This is really important when there are negative numbers involved.

Remember **BIDMAS** for the correct order of operations. You need to do:

Indices → Multiplication → Subtraction

Don't try to do more than one operation on each line of working.

Worked example

The area of this shape is A cm^2.

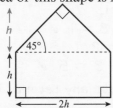

Everything in red is part of the answer.

Write a formula for A in terms of h.
Give your answer in its simplest form. **(3 marks)**

$A = 2h \times h + \frac{1}{2} \times 2h \times h$

$\quad = 2h^2 + h^2$

$A = 3h^2$

Problem solved!

You are only given the dimensions of the rectangle. You need to **infer** the height of the triangle. The angle of the slope is 45° and the base of the triangle is $2h$ so the height of the triangle must be h. Write this dimension on your diagram.

Use these dimensions to write an expression for the area of the triangle. Remember to write '$A =$'. If you only write '$3h^2$' it is an **expression**, not a **formula**.

You will need to use problem-solving skills throughout your exam – **be prepared!**

Now try this

The perimeter of this shape can be calculated using the formula

$$P = \frac{4(a^2 + ab + b^2)}{a + b}$$

Find the value of P when $a = 4.3$ cm and $b = 2.9$ cm.
Give your answer correct to 3 significant figures.

(2 marks)

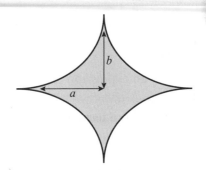

Arithmetic sequences

An **arithmetic or linear sequence** is a sequence of numbers where the difference between consecutive terms is **constant**. In your exam, you might need to work out the nth term of a sequence. Look at this example which shows you how to do it in four steps.

Worked example

1 Here is a sequence.

$1 \; \boxed{+4} \; 5 \; \boxed{+4} \; 9 \; \boxed{+4} \; 13 \; \boxed{+4} \; 17$

Work out a formula for the nth term of the sequence. **(2 marks)**

> Write in the difference between each term.

2 Here is a sequence.

Zero term

$-3 \quad 1 \; \boxed{+4} \; 5 \; \boxed{+4} \; 9 \; \boxed{+4} \; 13 \; \boxed{+4} \; 17$

Work out a formula for the nth term of the sequence. **(2 marks)**

> Work backwards to find the **zero term** of the sequence. You need to subtract 4 from the first term.

3 Here is a sequence.

Zero term

$-3 \quad 1 \; \boxed{+4} \; 5 \; \boxed{+4} \; 9 \; \boxed{+4} \; 13 \; \boxed{+4} \; 17$

Work out a formula for the nth term of the sequence. **(2 marks)**

nth term $=$ difference $\times \; n +$ zero term

> Write down the formula for the nth term. **Remember** this formula for the exam.

4 Here is a sequence.

Zero term

$-3 \quad 1 \; \boxed{+4} \; 5 \; \boxed{+4} \; 9 \; \boxed{+4} \; 13 \; \boxed{+4} \; 17$

Work out a formula for the nth term of the sequence. **(2 marks)**

nth term $=$ difference $\times \; n +$ zero term
nth term $= 4n - 3$

Is 99 in this sequence?

You can use the nth term to check whether a number is a term in the sequence.

The value of n in your nth term has to be a **positive** whole number.

Try some different values of n:

$n = 25 \rightarrow 4n - 3 = 97$
$n = 26 \rightarrow 4n - 3 = 101$

You can't use a value of n between 25 and 26 so 99 is **not** a term in the sequence.

Check it!

Check your answer by substituting values of n into your nth term.
1st term: when $n = 1$,
$4n - 3 = 4 \times 1 - 3 = 1 \; \checkmark$
2nd term: when $n = 2$,
$4n - 3 = 4 \times 2 - 3 = 5 \; \checkmark$

You can also generate any term of the sequence.

For the 20th term: when $n = 20$,
$4n - 3 = 4 \times 20 - 3 = 77$
So the 20th term is 77.

Now try this

Here are the first five terms of an arithmetic sequence.
3, 7, 11, 15, 19

(a) Write down an expression for the nth term. **(2 marks)**

Karen says that 89 is a term in the sequence.

(b) Is she right? Give reasons for your answer. **(2 marks)**

Method 1
Write down some terms in the sequence that are close to 89.

Method 2
Set the nth term equal to 89 and solve the equation.

Worked solution video

Solving sequence problems

Generating sequences

You can work out the terms of a sequence by substituting the term number into the nth term. Here are two examples:

nth term	$n^2 + 10$	$\sqrt{3}^n$
1st term	$1^2 + 10 = 11$	$\sqrt{3}^1 = \sqrt{3}$
2nd term	$2^2 + 10 = 14$	$\sqrt{3}^2 = 3$
3rd term	$3^2 + 10 = 19$	$\sqrt{3}^3 = 3\sqrt{3}$
⋮	⋮	⋮
8th term	$8^2 + 10 = 74$	$\sqrt{3}^8 = 81$

Sequences and equations

You can use the nth term or the term-to-term rule of a sequence to write an **equation**.

This sequence has term-to-term rule 'multiply by 2 then add a':

$$\boxed{\times 2} \boxed{+a} \quad \boxed{\times 2} \boxed{+a}$$

...   11   25   53   ...

So $2 \times 11 + a = 25$ and $a = 3$. You could use this information to find the next term in the sequence.

Worked example

The rule for finding the next term in a sequence is

$$\boxed{\text{Subtract } k \text{ then multiply by 3}}$$

The second term is 12 and the third term is 24.

...   12   24   ...

Work out the 1st term of the sequence.

(4 marks)

$24 = 3(12 - k)$

$24 = 36 - 3k$

$3k = 12$

$k = 4$

If the first term is T then:

$3(T - 4) = 12$

$3T - 12 = 12$

$3T = 24$

$T = 8$

Problem solved!

Work out what information you need to solve the problem. You can't find the 1st term until you know the value of k. You know two consecutive terms so you can solve an equation to find the value of k.

> You will need to use problem-solving skills throughout your exam – **be prepared!**

Fibonacci sequences

The rule for generating this sequence is 'add two consecutive terms to get the next term':

2, 3, 5, 8, 13, 21, ...

This type of sequence is called a Fibonacci sequence.

You can write the nth term of a sequence as u_n. Then the rule for this sequence would be:

$u_n + u_{(n+1)} = u_{(n+2)}$

Now try this

1 The nth term of a sequence is $50 - n^2$.
   (a) Write down the first three terms of the sequence. **(1 mark)**
   (b) Work out the first term of this sequence that is negative. **(1 mark)**

2 The rule for finding the next term in a sequence is

$$\boxed{\text{Add } k \text{ and then multiply by 2}}$$

The first four terms are a, b, 14, 46, ...
Work out the values of a and b. **(3 marks)**

3 The nth term of an arithmetic sequence is given by $u_n = 3n - 1$ where n is an integer.
   (a) Determine whether 52 is a term in this sequence. **(2 marks)**
   (b) Write an expression for u_n in terms of u_{n-1}. **(1 mark)**
   (c) The sum of two consecutive terms in this sequence is 55. Find the smaller of these two terms. **(3 marks)**

Quadratic sequences

If the nth term of a sequence contains an n^2 term and no higher power of n, it is called a quadratic sequence. You can write the nth term of a quadratic sequence as:

$$u_n = an^2 + bn + c$$

where a, b and c are numbers and a is not 0.

You need to be able to find the nth term of a quadratic sequence. You can use the golden rule on the right to help.

Make sure you are confident with arithmetic sequences on page 22 before tackling this page.

Golden rule

The **second differences** of a quadratic sequence are constant. The quadratic sequence with nth term $u_n = an^2 + bn + c$ has second differences equal to $2a$.

Here is the sequence $u_n = 3n^2 - n$

$$2 \quad 10 \quad 24 \quad 44 \quad 70 \quad 102 \ldots$$

$$+8 \quad +14 \quad +20 \quad +26 \quad +32$$

$$+6 \quad +6 \quad +6 \quad +6$$

The second differences are constant and are equal to $2a = 2 \times 3 = 6$.

Worked example

The diagram shows a sequence of patterns made from coins.

Pattern 1   Pattern 2   Pattern 3   Pattern 4   Pattern 5

The number of coins in each pattern forms a quadratic sequence. Find an expression, in terms of n, for the number of coins in the pattern n. **(4 marks)**

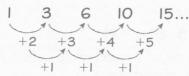

$$1 \quad 3 \quad 6 \quad 10 \quad 15 \ldots$$
$$+2 \quad +3 \quad +4 \quad +5$$
$$+1 \quad +1 \quad +1$$

$2a = 1$ so $a = 0.5$

$u_n = 0.5n^2 + bn + c$

n	1	2	3	4	5
u_n	1	3	6	10	15
$0.5n^2$	0.5	2	4.5	8	12.5
$u_n - 0.5n^2$	0.5	1	1.5	2	2.5

$bn + c = 0.5n$ so $b = 0.5$ and $c = 0$

$u_n = 0.5n^2 + 0.5n$

Pattern n contains $0.5n^2 + 0.5n$ coins

Start by writing out the number of coins in each pattern as a number sequence. You are told the sequence is quadratic so you know the **second differences** will be **constant**. The coefficient of n^2 in the nth term is **half the second difference**. The second differences are $+1$, so the value of a is 0.5 or $\frac{1}{2}$.

Once you have worked out the value of a, draw a table like this one. You need to compare the values of the terms u_n with $0.5n^2$. This will help you find the rest of the nth term. Add a row for $u_n - an^2$. This row will form an **arithmetic sequence** with nth term $bn + c$.

The arithmetic sequence 0.5, 1, 1.5, 2, 2.5... has nth term $u_n = 0.5n$. This is the last part of the nth term of your quadratic sequence.

Check it!

$n = 3$:

$0.5n^2 + 0.5n = 0.5 \times 3^2 + 0.5 \times 3$
$= 4.5 + 1.5 = 6$ ✓

Now try this

The first five terms of a quadratic sequence are 5, 9, 17, 29, 45, ...
Work out the nth term of this sequence.

(3 marks)

Straight-line graphs 1

Here are two things you need to know about straight-line graphs.

 If an equation is in the form $y = mx + c$, its graph will be a straight line.

$$y = -\tfrac{1}{2}x + 5$$

This number tells you the gradient of the graph.

The y-intercept of the graph is at (0, 5).

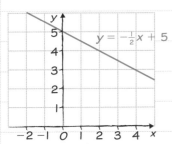

$y = -\tfrac{1}{2}x + 5$

The gradient of the graph is $-\tfrac{1}{2}$.

This means that for every unit you go across, you go half a unit down.

 Use a table of values to draw a graph.

$y = 2x + 1$

x	−1	0	1	2
y	−1	1	3	5

$y = 2 \times 2 + 1 = 5$

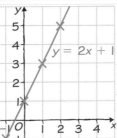

$y = 2x + 1$

Choose simple values of x and substitute them into the equation to find the values of y.

Plot the points on your graph and join them with a straight line.

Worked example

On the grid draw the graph of $x + y = 4$ for values of x from −2 to 5

(3 marks)

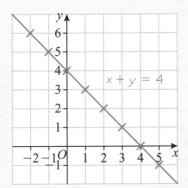

$x + y = 4$

 Everything in red is part of the answer.

x	−2	−1	0	1	2	3	4	5
y	6	5	4	3	2	1	0	−1

Finding equations

If you have a graph you can find its equation by working out the gradient and looking at the y-intercept.

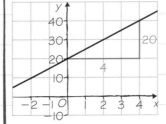

20

4

Draw a triangle to find the gradient.

Gradient $= \dfrac{20}{4} = 5$

The y-intercept is (0, 20).

Put your values for gradient, m, and y-intercept, c, into the equation of a straight line, $y = mx + c$.

The equation is $y = 5x + 20$

You can rearrange the equation of this graph into the form $y = mx + c$ so it is a **straight line**.

$x + y = 4$ $(-x)$

$y = -x + 4$ $m = -1$ and $c = 4$

The gradient is −1 and the y-intercept is at (0, 4). You could use this information to draw the graph, but it's safer to make a table of values. Make sure you plot **at least three** points, then join them with a straight line **using a ruler**.

Now try this

Find the equation of the straight line.

(3 marks)

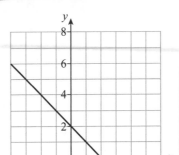

Use $y = mx + c$. Draw a triangle to find the gradient, m. The graph slopes down so m will be **negative**.

Straight-line graphs 2

On page 25 you revised how to find an equation by looking at a graph. You can also use **algebra** to find the equation. This is usually a more reliable and accurate method.

1 Given one point and the gradient

Substitute the gradient for m in $y = mx + c$
↓
Substitute the x- and y-values given into the equation
↓
Solve the equation to find c
↓
Write out the equation

Gradient 2, passing through point (3, 7)

$y = 2x + c$
$7 = 2 \times 3 + c$
$7 = 6 + c \quad (-6)$
$c = 1$
$y = 2x + 1$

2 Given two points

Draw a sketch showing the two points
↓
Work out the gradient of the line using a triangle
↓
Use method I (on the left) and one of the points given to find the equation

Passing through points (8, 20) and (10, 30)

Gradient $= \dfrac{10}{2} = 5$

$y = 5x + c$
$30 = 5 \times 10 + c$ so $c = -20$

$y = 5x - 20$

Worked example

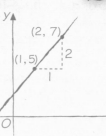

A line passes through the points with coordinates (1, 5) and (2, 7). Find the equation of the line.

(3 marks)

Gradient, $m = \dfrac{2}{1} = 2$
Equation of line: $y = mx + c \rightarrow y = 2x + c$
For point (1, 5), $x = 1$, $y = 5$
Substitute these values into the equation:
$5 = 2 + c \rightarrow c = 3$
The equation is $y = 2x + 3$

Positive or negative

If the line slopes **down** then the gradient is **negative**.

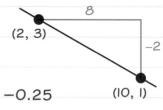

Gradient $= \dfrac{-2}{8} = -0.25$

Follow these steps:
1. Draw a **sketch** of the straight line.
2. Draw a **triangle** to find the gradient. This is your value for m. Put it into the equation for the line.
3. Use the x- and y-values of one of the points on the line to write an equation.
4. **Solve** your equation to find c.

Now try this

Worked solution video

1 A straight line has gradient 6 and passes through the point (5, 10).
Find an equation of the line. **(2 marks)**

2 Find the equation of a straight line which passes through the points (3, 5) and (7, 13). **(3 marks)**

3 A straight line passes through the points with coordinates $(-3, -2)$, $(1, 6)$ and $(k, 16)$.
Work out the value of k. You must show all your working. **(4 marks)**

Find the gradient, m, then substitute $x = 3$ and $y = 5$ into $y = mx + c$. Solve an equation to find the value of c.

Parallel and perpendicular

Parallel lines have the same gradient.

These three lines all have a gradient of 1

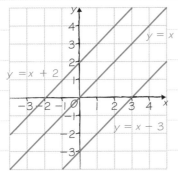

Perpendicular means at right angles.

If a line has gradient m then any line perpendicular to it will have gradient $-\dfrac{1}{m}$

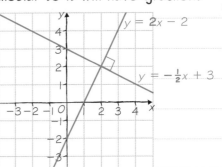

Worked example

A line L passes through the points $(-3, 6)$ and $(5, 4)$.

Another line, P, is perpendicular to L and passes through the point $(0, -7)$. Find the equation of line P. **(3 marks)**

Gradient of line L
$$= \frac{-2}{8} = \frac{-1}{4}$$

$(-3, 6)$ 8   L -2 $(5, 4)$

Gradient of line P
$$= \frac{-1}{\frac{-1}{4}} = 4$$

P passes through $(0, -7)$
Equation of P is: $y = 4x - 7$

1. Draw a sketch to find the gradient of line L.
2. The line slopes down so the gradient is negative.
3. Use $-\dfrac{1}{m}$ to calculate the gradient of P. If m is a fraction, you can just find its reciprocal and change the sign.
4. You know P passes through $(0, -7)$. Use $m = 4$ and $c = -7$ to write the equation of line P.

Check it!

If two lines are perpendicular the product of their gradients is -1: $-\dfrac{1}{4} \times 4 = -1$ ✔

Midpoints

A **line segment** is a short section of a straight line.

You can find the **midpoint** of a line segment if you know the coordinates of the ends.

$$\frac{-3 + 5}{2} \qquad \frac{8 + (-3)}{2}$$

$(-3, 8)$✗

$(1, 2\frac{1}{2})$

Midpoint ✗

✗$(5, -3)$

Coordinates of midpoint = (average of x-coordinates, average of y-coordinates)

Now try this

1   Line A has equation $y = \frac{1}{2}x + 1$.

Line B is parallel to line A.

Work out an equation for line B. **(3 marks)**

2   Point C has coordinates $(3, 4)$ and point D has coordinates $(9, 12)$.
    (a) Find the equation of the line CD. **(2 marks)**
    (b) Find the perpendicular bisector of the line CD. **(2 marks)**

Line B   P

$y = \frac{1}{2}x + 1$

Q

Line A

O   6 x

Quadratic graphs

Quadratic equations contain an x^2 term. Quadratic equations have **curved** graphs. You can draw the graph of a quadratic equation by completing a table of values.

> The **turning point** is the point where the direction of the curve changes.

Worked example

(a) Complete the table of values for $y = 4x - x^2$. **(2 marks)**

x	−1	0	1	2	3	4	5
y	−5	0	3	4	3	0	−5

(b) On the grid, draw the graph of $y = 4x - x^2$. **(2 marks)**
(c) Write down the coordinates of the turning point.

(1 mark)

(2, 4)

When $x = -1$: $4 \times (-1) - (-1)^2 = -4 - 1 = -5$
When $x = 4$: $4 \times 4 - 4^2 = 16 - 16 = 0$
Plot your points carefully on the graph and join them with a **smooth** curve.

Check it!
All the points on your graph should lie on the curve. If one of the points doesn't fit then double check your calculation.

> Everything in red is part of the answer.

Drawing quadratic curves

- ✓ Use a sharp pencil.
- ✓ Plot the points carefully.
- ✓ Draw a smooth curve that passes through every point.
- ✓ Label your graph.
- ✓ Shape of graph will be either ⋃ or ⋂

Drawing a smooth curve

It's easier to draw a smooth curve if you turn your graph paper so your hand is **inside** the curve.

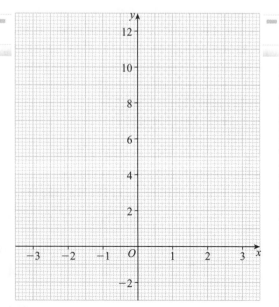

Now try this

Worked solution video

(a) Complete this table of values for $y = x^2 + 2$

x	−3	−2	−1	0	1	2	3
y		6		2	3	6	

(2 marks)

(b) Draw the graph of $y = x^2 + 2$ for
$x = -3$ to $x = 3$ **(2 marks)**

(c) Use your graph to find the value of
y when $x = 2.5$ **(1 mark)**

Cubic and reciprocal graphs

You might need to **draw** or **interpret** cubic and reciprocal graphs in your exam. You can use a **table of values** to draw any graph, but it helps if you know what the **general shape** of the graph is going to be.

1 Cubic graphs

Graphs that contain an x^3 term and no higher powers of x are called **cubic graphs**. Here are two examples.

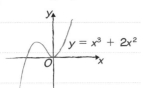

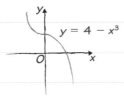

2 Reciprocal graphs

Graphs of the form $y = \frac{k}{x}$ where k is a number are called **reciprocal graphs**.

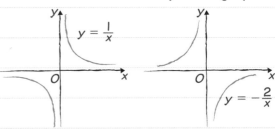

The graphs get closer and closer to the x- and y-axes but never touch them.

Worked example

(a) Complete the table of values for $y = x^3 - 4x - 3$

x	-2	-1	0	1	2	3
y	-3	0	-3	-6	-3	12

(2 marks)

(b) On the grid, draw the graph of $y = x^3 - 4x - 3$ for $-2 \leqslant x \leqslant 3$ **(2 marks)**

(c) (i) Estimate the value of x when $y = 6$ **(1 mark)**

2.7

(ii) Comment on the accuracy of your estimate. **(1 mark)**

The estimate is not very accurate because it is based on reading off a graph.

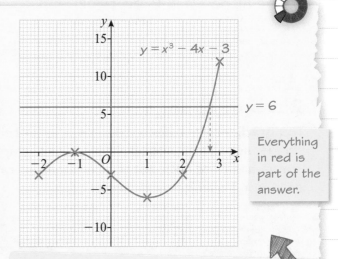

Everything in red is part of the answer.

This is a **cubic graph** with a **positive** coefficient of x. If you recognise the shape of the graph then it's easier to tell if you've plotted your coordinates correctly.

Now try this

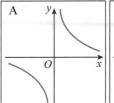
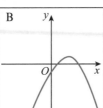
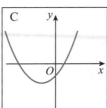
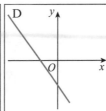
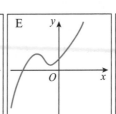
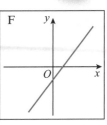

Write down the letter of the graph which could have the equation

(a) $y = -4x - 2$

(b) $y = x^3 - 3x + 4$

(c) $y = \frac{1}{x}$

(d) $y = 8x - 15 - x^2$

(e) $y = 2x - 3$

(f) $y = x^2 + 2x - 8$ **(6 marks)**

Real-life graphs

Distance–time graphs

A **distance–time** graph shows how distance changes with time. This distance–time graph shows Jodi's run. The shape of the graph gives you information about the journey.

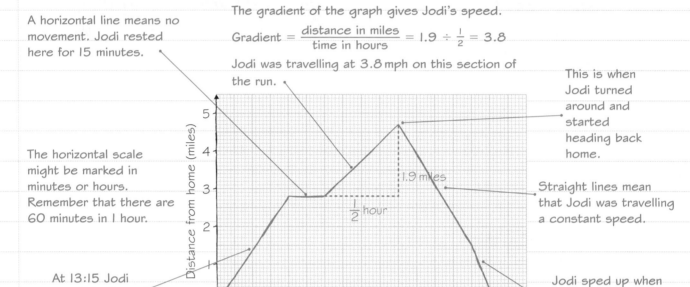

A horizontal line means no movement. Jodi rested here for 15 minutes.

The gradient of the graph gives Jodi's speed.

$$\text{Gradient} = \frac{\text{distance in miles}}{\text{time in hours}} = 1.9 \div \frac{1}{2} = 3.8$$

Jodi was travelling at 3.8 mph on this section of the run.

The horizontal scale might be marked in minutes or hours. Remember that there are 60 minutes in 1 hour.

This is when Jodi turned around and started heading back home.

1.9 miles

$\frac{1}{2}$ hour

Straight lines mean that Jodi was travelling a constant speed.

At 13:15 Jodi was 1.4 miles from home.

Jodi sped up when she was nearly home. The graph is steeper here.

Rates of change

The **gradient** on a distance–time graphs tells you the **rate of change** of distance with time. This is also called **speed**. You can use graphs to find other rates of change. These garden ponds are filled with water at a constant rate. The graphs show how the depth of water in each pond changes with time.

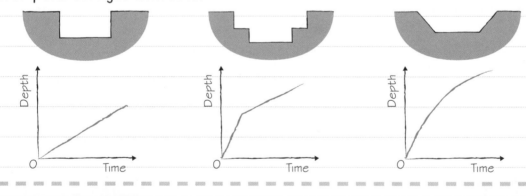

The gradient of the graph tells you the **rate of change** of the depth of water at that point. Where the pond is narrower, the rate of change will be higher.

Worked solution video

Now try this

Christina rode her bike to a friend's house. She rested once on the way. She had coffee at her friend's house, then rode home.

(a) How long did Christina spend at her friend's house? **(1 mark)**

(b) Work out Christina's average speed for her return journey. **(2 marks)**

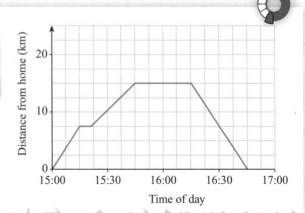

Quadratic equations

Quadratic equations can be written in the form $ax^2 + bx + c = 0$ where a, b and c are numbers.

You need to be able to **solve** a quadratic equation without using your calculator.

1. **Rearrange** it into the form
   $ax^2 + bx + c = 0$

2. **Factorise** the left-hand side.

3. Set each factor **equal to zero** and solve to find two values of x.

For a reminder about factorising quadratic expressions have a look at page 18.

Two to watch

1 When $c = 0$:
$x^2 - 10x = 0$
$x(x - 10) = 0$
Solutions are $x = 0$ and $x = 10$

2 When $b = 0$ (difference of two squares):
$9x^2 - 4 \qquad = 0$
$(3x + 2)(3x - 2) = 0$
Solutions are $x = \frac{2}{3}$ and $x = -\frac{2}{3}$

Worked example

Solve $x^2 + 8x - 9 = 0$ **(3 marks)**

$(x + 9)(x - 1) = 0$

$x + 9 = 0 \qquad x - 1 = 0$
$x = -9 \qquad x = 1$

Follow the three steps given above.

1. The equation is already in the right form.

2. To factorise look for two numbers which add up to 8 and multiply to make -9. The numbers are 9 and -1.

3. Set each factor equal to 0 and solve.

Check it!
$(1)^2 + 8(1) - 9 = 1 + 8 - 9 = 0$ ✓
$(-9)^2 + 8(-9) - 9 = 81 - 72 - 9 = 0$ ✓

Worked example

Solve $2(x + 1)^2 = 3x + 5$ **(4 marks)**

$2(x^2 + 2x + 1) = 3x + 5$
$2x^2 + 4x + 2 = 3x + 5$
$2x^2 + x - 3 = 0$
$(2x + 3)(x - 1) = 0$

$2x + 3 = 0 \qquad x - 1 = 0$
$x = -\frac{3}{2} \qquad x = 1$

When you are solving a quadratic equation you must rearrange it into the form $ax^2 + bx + c = 0$ **before** you factorise.

Be really careful if the coefficient of x^2 is bigger than 1. The two factors will look like this:

$2x^2 + x - 3 = 0$
$(2x \pm \ldots)(x \pm \ldots) = 0$

The number part of the expression is -3, so the numbers in the factors must be either -1 and 3 or 1 and -3.

Now try this

When you are solving an equation you must always show **clear algebraic working**. Don't just write down the answers without any working.

Solve
 (a) $m^2 - 8m + 12 = 0$ **(3 marks)**

 (b) $w^2 - 36 = 5w$ **(3 marks)**

 (c) $5y^2 + 37y - 24 = 0$ **(3 marks)**

(d) $8x^2 - 4 = (x - 1)^2$ **(4 marks)**

The quadratic formula

The solutions of the quadratic equation
$ax^2 + bx + c = 0$
where $a \neq 0$ are given by

$$x = \frac{-b \pm \sqrt{b^2 - 4ac}}{2a}$$

You can use this formula for **any** quadratic equation. But be careful, not all quadratic equations have a solution.

LEARN IT!

Safe substituting

Equation is in the form
$ax^2 + bx + c = 0$ ✓

Write down your values of a, b and c before you substitute. ✓

Use brackets when you are substituting negative numbers. ✓

Show what you have substituted in the formula. ✓

Simplify what is under the square root and write this down. ✓

The $\pm$ symbol means you need to do two calculations. ✓

Worked example

Solve $5x^2 + x + 11 = 14$

Give your solutions correct to 3 significant figures.

Show your working clearly.   **(3 marks)**

$5x^2 + x - 3 = 0$
$a = 5, b = 1, c = -3$

$$x = \frac{-1 \pm \sqrt{1^2 - 4 \times 5 \times (-3)}}{2 \times 5}$$

$$= \frac{-1 + \sqrt{61}}{10} \text{ or } \frac{-1 - \sqrt{61}}{10}$$

$$= 0.681024... \text{ or } -0.881024...$$

$$= 0.681 \text{ or } -0.881 \text{ (to 3 s.f.)}$$

You are asked to find 'solution**s**'. This tells you that you are solving a quadratic equation.

You must give your answer 'correct to 3 significant figures'. This tells you that you need to use the quadratic formula.

Write down at least five figures after the decimal point on the calculator display before giving your final answer. You might need to use the S⇔D button on your calculator to get your answer as a decimal.

How many solutions?

A quadratic equation can have two solutions, one solution or no solutions.
You can use $b^2 - 4ac$ (the part under the square root) to work out how many solutions a quadratic equation has.

If $b^2 - 4ac$ is negative, there are no solutions.

You can't calculate the square root of a negative number.

If $b^2 - 4ac = 0$ there is only one solution.

± 0 appears in the formula, so you get the same answer whether you use + or −

If $b^2 - 4ac > 0$ there are two different solutions.

Now try this

1 Solve, giving your solutions correct to 2 decimal places.

  (a) $7x^2 + 3x - 6 = 0$ **(2 marks)**
  (b) $m^2 + 50m = 6000$ **(3 marks)**

2 Solve this quadratic equation.
  $2x(x - 5) = 3x - 1$
  Give your solutions correct to 3 significant figures.   **(3 marks)**

Worked solution video

Completing the square

If a quadratic expression is written in the form $(x + p)^2 + q$ it is in **completed square** form. You can solve quadratic equations which don't have integer answers by completing the square.

Useful identities

If you learn these two identities you can save time when you are completing the square.

 $x^2 + 2bx + c \equiv (x + b)^2 - b^2 + c$

 $x^2 - 2bx + c \equiv (x - b)^2 - b^2 + c$

There is more about identities on page 52.

Positive and negative roots

Remember that any positive number has **two** square roots: one **positive** and one **negative**.

If you 'square root' both sides of an equation you need to use ± (plus-or-minus) to show that there are two square roots.

$$x^2 = 4 \qquad\qquad (x + 4)^2 = 3$$
$$x = \pm 2 \qquad\qquad (x + 4) = \pm\sqrt{3}$$
$$\qquad\qquad\qquad x = -4 \pm\sqrt{3}$$

Worked example

(a) Find values of p and q such that
$x^2 + 6x - 20 \equiv (x + p)^2 + q$ **(2 marks)**

$x^2 + 6x - 20 = (x + 3)^2 - 3^2 - 20$
$\qquad\qquad = (x + 3)^2 - 9 - 20$
$\qquad\qquad = (x + 3)^2 - 29$
$\qquad\qquad p = 3$ and $q = -29$

(b) Hence, or otherwise, solve the equation
$x^2 + 6x - 20 = 0$

Give your answer in surd form.   **(2 marks)**

$(x + 3)^2 - 29 = 0 \qquad (+29)$
$(x + 3)^2 = 29 \qquad (\sqrt{\ })$
$x + 3 = \pm\sqrt{29} \quad (-3)$
$x = -3 \pm\sqrt{29}$

Compare the expression with the identities for completing the square.
$x^2 + 6x - 20 \equiv (x + p)^2 + q$
$x^2 + 2bx + c \equiv (x + b)^2 - b^2 + c$
$2b = 6$ so $b = 3$
$c = -20$

Substitute these values into the identity and simplify to find p and q.

Use your answer to part (a) to write the expression in completed square form. The unknown only appears once so you can solve it using **inverse operations**.

Remember to use the ± symbol when you take square roots of both sides. The two solutions are $x = -3 + \sqrt{29}$ and $x = -3 - \sqrt{29}$

Now try this

 1 Solve, giving your answers in surd form
   (a) $x^2 + 10x + 12 = 0$ **(4 marks)**
   (b) $y^2 - 6y - 15 = 0$ **(4 marks)**

 2 Write $2x^2 + 20x + 7$ in the form
$a(x + p)^2 + q$ **(3 marks)**

  Worked solution video

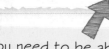 You need to be able to do these **without** a calculator.

You can start by writing the expression as $2(x^2 + 10x) + 7$, then writing $x^2 + 10x$ in completed square form.

Simultaneous equations 1

Simultaneous equations have two unknowns. You need to find the values for the two unknowns which make **both** equations true.

Algebraic solution

1. Number each equation.

2. If necessary, multiply the equations so that the coefficients of one unknown are the same.

3. Add or subtract the equations to **eliminate** that unknown.

4. Once one unknown is found use substitution to find the other.

5. Check the answer by substituting both values into the other equation.

$$3x + y = 20 \quad ①$$
$$x + 4y = 14 \quad ②$$

$$12x + 4y = 80 \quad ① \times 4$$
$$- (x + 4y = 14) \quad -②$$
$$\overline{}$$
$$11x = 66$$
$$x = 6$$

Substitute $x = 6$ into ①:
$$3(6) + y = 20$$
$$18 + y = 20$$
$$y = 2$$

Solution is $x = 6$, $y = 2$.
Check it: $x + 4y = 6 + 4(2) = 14$ ✓

Worked example

Solve the simultaneous equations
$$6x + 2y = -3 \quad ①$$
$$4x - 3y = 11 \quad ② \quad \textbf{(4 marks)}$$

$$18x + 6y = -9 \quad ① \times 3$$
$$+ \; 8x - 6y = 22 \quad ② \times 2$$
$$\overline{}$$
$$26x = 13$$
$$x = \tfrac{1}{2}$$

Substitute $x = \tfrac{1}{2}$ into ①:
$$6\left(\tfrac{1}{2}\right) + 2y = -3$$
$$3 + 2y = -3$$
$$2y = -6$$
$$y = -3$$

Easier eliminations

You can save time by choosing the right unknown to eliminate. Look for one of these:

1 If an unknown appears **on its own** in one equation you only need to multiply one equation to eliminate that unknown.

2 If an unknown has **different signs** in the two equations you can eliminate by **adding**.

Multiply both equations by a whole number to make the coefficients of y the same.

Check it!

Always use the equation you **didn't** substitute into to check your answer:
$$4x - 3y = 4\left(\tfrac{1}{2}\right) - 3(-3) = 2 + 9 = 11 \checkmark$$

Graphical solution

You can solve these simultaneous equations by drawing a graph.
$$x - y = 1 \qquad x + 2y = 4$$
The coordinates of the point of intersection give the solution to the simultaneous equations.
The solution is $x = 2$, $y = 1$

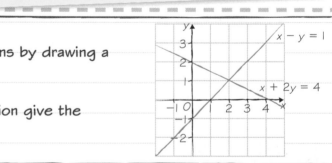

Now try this

1 Solve the simultaneous equations
$$3x - 2y = 12$$
$$x + 4y = 11 \qquad \textbf{(3 marks)}$$

Draw a coordinate grid from −3 to 5 in both directions.

2 By drawing two suitable straight lines on a coordinate grid, solve these simultaneous equations.
$$3y + 2x = 6$$
$$y = 2x - 2 \qquad \textbf{(4 marks)}$$

Worked solution video

Simultaneous equations 2

If a pair of simultaneous equations involves an x^2 or a y^2 term, you need to solve them using **substitution**. Remember to **number** the equations to keep track of your working.

Rearrange the linear equation to make y the subject.

$$y = x^2 - 2x - 7 \qquad ①$$
$$x - y = -3 \qquad ②$$
$$y = x + 3 \qquad ③$$
$$x + 3 = x^2 - 2x - 7 \qquad \text{Substitute ③ into ①.}$$
$$0 = x^2 - 3x - 10$$
$$0 = (x - 5)(x + 2)$$
$$x = 5 \text{ or } x = -2$$

Each solution for x has a corresponding value of y. Substitute into (3) to find the two solutions.

The solutions are $x = 5$, $y = 8$ and $x = -2$, $y = 1$

Worked example

Solve the simultaneous equations

$x - 2y = 1 \qquad ①$

$x^2 + y^2 = 13 \qquad ②$ **(5 marks)**

$x = 1 + 2y \qquad ③$

Substitute ③ into ②:

$(1 + 2y)^2 + y^2 = 13$

$1 + 4y + 4y^2 + y^2 = 13$

$5y^2 + 4y - 12 = 0$

$(5y - 6)(y + 2) = 0$

$y = \dfrac{6}{5} \qquad \text{or} \quad y = -2$

$x = 1 + 2\left(\dfrac{6}{5}\right) \qquad x = 1 + 2(-2)$

$\quad = \dfrac{17}{5} \qquad\qquad\quad = -3$

Solutions: $x = \dfrac{17}{5}$, $y = \dfrac{6}{5}$ and

$\qquad\qquad x = -3$, $y = -2$

This equation has **two pairs** of solutions. Each solution is an x-value **and** a y-value. You need to find four values in total, and pair them up correctly.

You can substitute for x or y. It is easier to substitute for x because there will be no fractions.

Use brackets to make sure that the whole expression is squared.

Rearrange the quadratic equation for y into the form $ay^2 + by + c = 0$

Factorise the left-hand side to find two solutions for y.

Substitute each value of y into equation ① to find the corresponding values of x.

Thinking graphically

The solutions to a simultaneous equation correspond to the points where the graphs of each equation **intersect**.

Because an equation involving x^2 or y^2 represents a **curve**, there can be two points of intersection. Each point has an x-value and a y-value. You can write the solutions using coordinates.

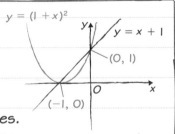
$y = (1 + x)^2$, $y = x + 1$, $(0, 1)$, $(-1, 0)$

Now try this

1 Solve the simultaneous equations

$x - 2y = 3$

$x^2 + 2y^2 = 27$ **(5 marks)**

2 L is the straight line with equation $y = 2x - 1$

C is the circle with equation $x^2 + y^2 = 13$

The line intersects the circle at two points.

Find the coordinates of both points.   **(5 marks)**

There is more about equations of circles on the next page.

Equation of a circle

A circle of radius r centred at the origin has equation $x^2 + y^2 = r^2$

LEARN IT!

Be careful with the right-hand side of this equation. It is the radius **squared**.

A line which just touches the circle once is called a **tangent** to the circle. The angle between the tangent and the radius is 90°.

There is more about circles and tangents on page 104.

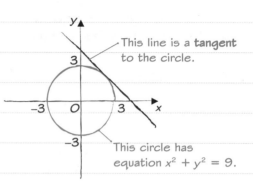

This line is a **tangent** to the circle.

This circle has equation $x^2 + y^2 = 9$.

Worked example

C is the circle with equation $x^2 + y^2 = 100$

(a) Show that the point A with coordinates (6, –8) lies on C. **(1 mark)**

At the point (6, –8), $x^2 + y^2 = 6^2 + (-8)^2$
$$= 36 + 64$$
$$= 100$$

So (6, –8) lies on the circle.

T is a tangent to the circle at the point (6, –8)

(b) Find the equation of **T**. **(5 marks)**

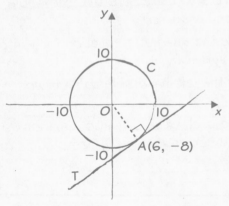

Gradient of $OA = -\frac{8}{6} = -\frac{4}{3}$

So gradient of $\mathbf{T} = \frac{3}{4}$

Equation of **T** is $y = \frac{3}{4}x + c$

T passes through (6, –8) so

$-8 = \frac{3}{4} \times 6 + c$

$c = -12.5$

So the equation of **T** is $y = \frac{3}{4}x - 12.5$

Problem solved!

Start by drawing a **sketch** so you can see what is going on. A sketch should be neat, so use a ruler and compasses. Label all the important information.

A tangent to a circle is always perpendicular to the radius of the circle. Use this fact to find the gradient of the tangent. The line OA is a radius of the circle.

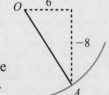

Because OA is perpendicular to **T**:

Gradient of $OA = \dfrac{-1}{\text{Gradient of } \mathbf{T}} = \dfrac{-1}{\left(-\frac{4}{3}\right)} = \dfrac{3}{4}$

For a reminder about gradients of perpendicular lines have a look at page 27.

Now you know the gradient of **T** and one point that it passes through. You can use the skills you revised on page 26 to find its equation.

You will need to use problem-solving skills throughout your exam – **be prepared!**

Now try this

On graph paper, draw axes from −8 to 8 in both directions.

(a) Draw the graph of $x^2 + y^2 = 16$. **(2 marks)**

(b) Draw the line of $y = \frac{1}{2}x + 6$ on the same axes. **(1 mark)**

(c) Use your graph to explain why there are no solutions to the simultaneous equations
$x^2 + y^2 = 16$
$2y - x = 12$ **(2 marks)**

Have a look at pages 34 and 35 for a reminder about simultaneous equations.

Inequalities

An inequality tells you when one value or expression is bigger or smaller than another value. You can represent **inequalities** on a number line.

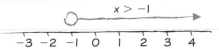

Use an **open** circle for $>$ and $<$

The open circle shows that -1 is **not** included.

Use a **closed** circle for $\geq$ and $\leq$

The closed circle shows that 3 **is** included.

Solving inequalities

You can solve an inequality in exactly the same way as you solve an equation.

$x - 3 \leq 12$ $(+ 3)$

     $x \leq 15$

The solution has the letter on its own on one side of the inequality and a number on the other side.

Golden rule

If you **multiply** or **divide** both sides of an inequality by a **negative** number you have to **reverse** the **inequality** sign.

$6 - 5x > 10$ $(- 6)$

   $-5x > 4$ $(\div -5)$

     $x < \dfrac{-4}{5}$

You have divided by a negative number so you have to reverse the inequality sign.

Worked example

Solve $8x - 7 > 3x + 3$ **(2 marks)**

               $(+ 7)$

$8x > 3x + 10$ $(- 3x)$

$5x > 10$ $(\div 5)$

$x > 2$

This is an **inequality** and not an equation. So don't use an '=' sign in your answer. The solution has the letter on its own on one side and a number on the other.

You could also use **set notation** to write your answer. You would write $\{x : x > 2\}$. This means 'the set of all values of x such that x is greater than 2'.

Integer solutions

You might need to write down all the integer solutions that **satisfy** an inequality.

Integers are positive or negative whole numbers, including 0.

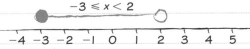

This shows that x is between -3 and 2. It can equal -3 but cannot equal 2.

The integer solutions that satisfy this inequality are $-3, -2, -1, 0$ and 1.

Now try this

1 Solve the inequality
  (a) $3n + 8 > 2$ **(2 marks)**
  (b) $2(n - 5) \geq n + 12$ **(2 marks)**

2 Find the integer value of x that satisfies both these inequalities.
  $x - 5 > 2$ $3x - 10 < 17$ **(3 marks)**

3 Find all the integer values of n that satisfy both these inequalities.
  $n + 5 > 2$ $4 - 2n \geq 1$ **(3 marks)**

4 Solve the inequality.
  $\dfrac{x + 1}{3} < \dfrac{1 - 2x}{2}$ **(3 marks)**

Quadratic inequalities

On page 37 you revised **linear inequalities**. In your exam, you might need to solve a **quadratic inequality**, which involves an x^2 term.

Using a sketch

This graph shows a **sketch** of the **curve** $y = x^2$, and the **straight line** $y = 36$

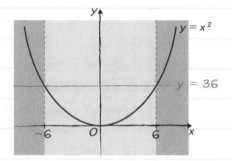

  For these values of x, the curve is **below** the line, so $x^2 < 36$

The solutions of $x^2 < 36$ are $-6 < x < 6$

  For these values of x, the curve is **above** the line, so $x^2 > 36$

The solutions of $x^2 > 36$ are $x < -6$ or $x > 6$

Worked example

Solve the inequality $m^2 \leqslant 25$ **(2 marks)**

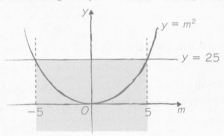

$-5 \leqslant m \leqslant 5$

Draw a sketch of $y = m^2$ and $y = 25$
$\sqrt{25} = 5$ so the graphs intersect at $m = 5$ and $m = -5$
You want m^2 to be **less than or equal to** 25, so you need to consider the values of m between -5 and 5. Make sure you use the same type of inequality signs as that given in the question.

Check it!

Choose a value in your solution range and check it satisfies the original inequality:
$(-3)^2 = 9 \leqslant 25$ ✓

Worked example

Solve the inequality $2x^2 - 5x - 3 > 0$. Represent your solution on this number line.

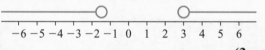

                             (3 marks)

$(2x + 1)(x - 3) > 0$

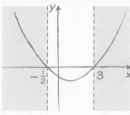

$x < -\dfrac{1}{2}$ or $x > 3$

Factorise the left-hand side then sketch the graph. The inequality is $>$ so you are interested in the x-values where the graph is **above** the horizontal axis.

You could also write this solution using **set notation** as $\{x : x > 3\} \cup \{x : x < -\frac{1}{2}\}$.

This symbol means 'union'. There's more about this on page 125.

Now try this

1 Solve the inequality $p^2 < 4$
Represent your solution on this number line.

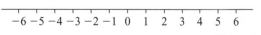

                       (2 marks)

2 Find the set of values of x for which
$x^2 - 5x - 14 < 0$ **(4 marks)**

Start by sketching the graph of $y = x^2 - 5x - 14$. You are interested in the values of x where the graph is **below** the horizontal axis.

Trigonometric graphs

You need to recognise and be able to sketch the graphs of the **trigonometric functions** sin, cos and tan. Revise the trigonometric functions on pages 77–79.

 1 $y = \sin x$ and $y = \cos x$

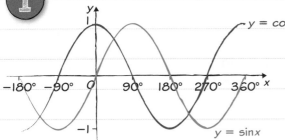

- $y = \sin x$ is the **same shape** as $y = \cos x$
- $y = \sin x$ is **translated** 90° to the right
- $y = \cos x$ is **symmetrical** about the y-axis
- $y = \sin x$ is **symmetrical** about the line $x = 90°$.

 2 $y = \tan x$

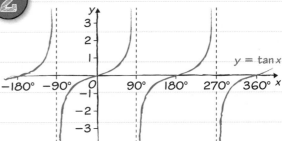

- $y = \tan x$ **repeats** every 180°
- There are **asymptotes** at –90°, 90°, 270°...
- The graph gets closer to these asymptotes but never reaches them.

Worked example

(a) Sketch the graph of $y = \tan x°$ for values of x from 0 to 360. **(3 marks)**

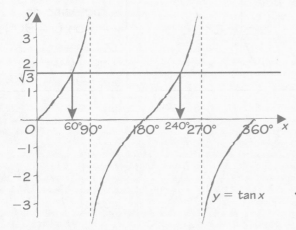

(b) $\tan 60° = \sqrt{3}$

Write down one other value of x that satisfies $\tan x° = \sqrt{3}$ **(2 marks)**

$270° – 30° = 240°$

Sketching a trig graph

✓ Label the x-axis in multiples of 90°

✓ For sin and cos label the y-axis from –1 to 1

✓ For tan label the y-axis from –3 to 3

✓ Mark some values that you know on your graph

For example $\cos 0° = 1$ and $\cos 90° = 0$. There is more about exact values of trig functions on page 79.

You can use the graph to find trig values for any angle. The graph of $y = \tan x$ **repeats** every 180°, so the value of $\tan x$ will be the same as the value of $\tan(270° – x)$. You can check this result using a calculator.

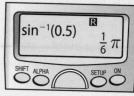

Now try this

(a) Sketch the graph of $y = \cos x°$ for values of x from 0 to 360. **(3 marks)**

(b) $\cos 45° = \dfrac{1}{\sqrt{2}}$. Write down one other value of x that satisfies $\cos x° = \dfrac{1}{\sqrt{2}}$. **(2 marks)**

Transforming graphs

You can change the equation of a graph to **translate** or **reflect** it. The easiest way to describe these transformations is using **function notation**. The tables show transformations of the graph $y = f(x)$. There's more on function notation on page 50.

Function	$y = f(x) + a$	$y = f(x + a)$
Transformation of graph	Translation $\begin{pmatrix} 0 \\ a \end{pmatrix}$	Translation $\begin{pmatrix} -a \\ 0 \end{pmatrix}$
Useful to know	$f(x) + a \rightarrow$ move **up** a units $f(x) - a \rightarrow$ move **down** a units	$f(x + a) \rightarrow$ move **left** a units $f(x - a) \rightarrow$ move **right** a units
Example	$y = f(x) + 3$ $y = f(x)$ 3	$y = f(x)$ $y = f(x + 5)$ −5

You can combine these translations. The graph of $y = f(x + 2) - 3$ would be a translation $\begin{pmatrix} -2 \\ -3 \end{pmatrix}$. Look at page 43 to see how you can use this fact to sketch quadratic graphs.

Function	$y = -f(x)$	$y = f(-x)$
Transformation of graph	Reflection in the x-axis	Reflection in the y-axis
Useful to know	'−' outside the bracket	'−' inside the bracket
Example	$y = f(x)$ $y = -f(x)$	$y = f(-x)$ $y = f(x)$

Sin and cos

The graph of $y = \sin x$ is a translation of $y = \cos x$ by 90° to the **right**. This means that $y = \sin x$ is the same as $y = \cos (x - 90°)$.

Have a look at the sketches of these graphs on page 39.

Worked example

The diagram shows a sketch of a curve with equation $y = f(x)$.

On the same diagram sketch the curve with equation
(a) $y = f(x + 3)$ **(3 marks)**
(b) $y = -f(x)$ **(3 marks)**

Show clearly the coordinates of any turning points, and any points of intersection with the axes.

Everything in red is part of the answer.

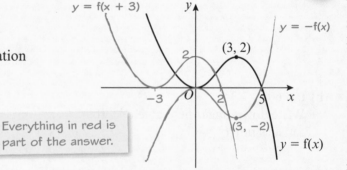

Now try this

The diagram shows a sketch of the curve with equation $y = f(x)$. It passes through the origin O and has a turning point at $(2, -4)$.

Write down the coordinates of the turning point on the curve with equation

(a) $y = f(x - 3)$ **(1 mark)**   (b) $y = f(x) - 5$ **(1 mark)**
(c) $y = -f(x)$ **(1 mark)**   (d) $y = f(-x)$ **(1 mark)**

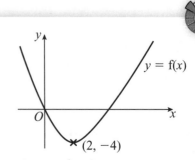

Inequalities on graphs

You can show the points that satisfy inequalities involving x and y on a graph.

For example, follow these steps to shade the region R that satisfies the inequalities

$$x \geqslant 2 \qquad y > x \qquad x + y \geqslant 6$$

Always work on one inequality at a time.

1 $x \geqslant 2$
Draw the graph of $x = 2$ with a solid line. Use a small arrow to show which side of the line you want.

2 $y > x$
Draw the graph of $y = x$ with a dotted line.
Show which side of the line you want.

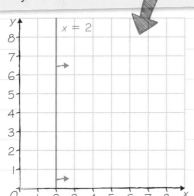

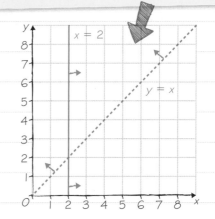

3 $x + y \geqslant 6$
Draw the graph of $x + y = 6$ with a solid line. Use a table of values.

x	0	3	6
y	6	3	0

Show which side of the line you want.
$x + y$ increases as you move away from the origin.
Shade in the region and label it **R**.

4 Check it!
Pick a point inside your shaded region. Check that the x- and y-values for that point satisfy **all** the inequalities.
At $(4, 5)$ $x = 4$ and $y = 5$.
$x \geqslant 2$ ✓
$y > x$ ✓
$x + y \geqslant 6$ ✓

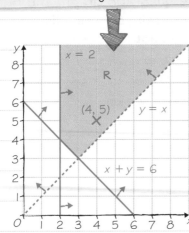

Graphical inequalities checklist

$<$ and $>$ are shown by **dotted** lines.

$\leqslant$ and $\geqslant$ are shown by **solid** lines.

Points on a solid line **are** included in the region.

Points on a dotted line **aren't** included in the region.

Now try this

Draw a coordinate grid from –2 to 8 in both directions. Shade the region of points whose coordinates satisfy these inequalities.

$x \geqslant -1 \qquad y > 5 \qquad y \geqslant x + 3$ **(4 marks)**

Worked solution video

Using quadratic graphs

You can use graphs to solve quadratic equations. You will need to look for the x-values of the points where a quadratic graph intersects with a straight line.

This grid shows one quadratic graph and two straight-line graphs.

The x-values at A and B are the solutions of the quadratic equation

$$x^2 - 4x = 12$$

or

$$x^2 - 4x - 12 = 0$$

The x-values at C and D are the solutions of the quadratic equation

$$x^2 - 4x = x + 3$$

or

$$x^2 - 5x - 3 = 0$$

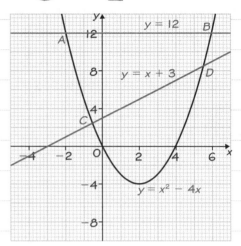

Worked example

This is a graph of $y = 2x^2 + 5x$

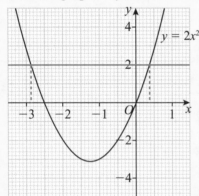

Everything in red is part of the answer.

By drawing a suitable straight line on the graph, find estimates for the solutions of the equation $2x^2 + 5x - 2 = 0$

Give your answers correct to 1 decimal place. **(3 marks)**

$$2x^2 + 5x - 2 = 0 \qquad (+ 2)$$
$$2x^2 + 5x = 2$$
$$x = 0.4, \; x = -2.9$$

You can solve the **quadratic equation** $2x^2 + 5x - 2 = 0$ by finding where the graph $y = 2x^2 + 5x$ crosses the straight line $y = 2$

Draw the line $y = 2$ on the graph.

The solutions are the x-values at the points of intersection.

A suitable line

To find a **suitable** straight line, rearrange the quadratic equation so that the left-hand side matches the equation of the quadratic graph. The right-hand side will tell you which line to draw. Here is another example:

Graph given:	$y = x^2 - 6x + 4$
Equation to solve:	$x^2 - 5x + 1 = 0$
Rearrange equation:	$x^2 - 6x + 4 = -x + 3$
Line to draw:	$y = -x + 3$

Now try this

(a) Complete the table of values for $y = x^2 + 3x - 3$

x	−5	−4	−3	−2	−1	0	1	2
y	7	1			−3	1	7	

(2 marks)

(b) On a grid with $-5 \leqslant x \leqslant 5$ and $-8 \leqslant y \leqslant 8$, draw the graph of $y = x^2 + 3x - 3$ **(2 marks)**

(c) Find estimates for the solutions of $x^2 + 3x - 3 = 0$ **(1 mark)**

(d) By drawing a suitable straight line, work out the solutions of the equation $x^2 + 2x - 4 = 0$ **(3 marks)**

Turning points

You can find the turning point of a quadratic graph by writing the function in **completed square** form. Make sure you are confident with completing the square before tackling this page. Have a look at page 33 for a reminder.

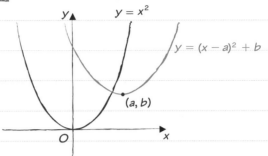

$y = x^2$

$y = (x - a)^2 + b$

(a, b)

Golden rule

The graph of
$y = (x - a)^2 + b$
has a turning point
at (a, b)

LEARN IT!

The graph of $y = (x - a)^2 + b$ is a translation of the graph of $y = x^2$ by the vector $\binom{a}{b}$. There is more on transformations of graphs on page 40.

Worked example

(a) $f(x) = x^2 + 3x + 5$
Sketch the graph of $y = f(x)$, showing the coordinates of the turning point and the coordinates of any intercepts with the coordinate axes. **(4 marks)**

$x^2 + 3x + 5 = (x + 1.5)^2 - 1.5^2 + 5$
$\qquad\qquad\qquad = (x + 1.5)^2 + 2.75$
so turning point at $(-1.5, 2.75)$
When $x = 0$, $f(x) = 5$ so graph
intercepts y-axis at $(0, 5)$

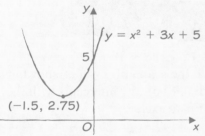

$y = x^2 + 3x + 5$

5

$(-1.5, 2.75)$

(b) Hence, or otherwise, determine whether $f(x)$ has any real roots. Give reasons for your answer. **(2 marks)**
$y = f(x)$ does not intercept x-axis, so $f(x)$ has no real roots.

Problem solved!

If a question asks you to **sketch** the graph of a quadratic function you should use **algebra** to find the turning point and any intercepts with the coordinate axes.
1. **Complete the square** to find the turning point.
2. Set $x = 0$ to work out the y-intercept.
3. Draw axes with a ruler, then sketch the graph.
4. Mark the coordinates of the turning point and the point where the graph crosses the y-axis, and label your graph.

Have a look at page 28 for a reminder about the shapes of quadratic graphs.

You will need to use problem-solving skills throughout your exam – **be prepared!**

Functions and roots

The roots of a function $f(x)$ are the values of x for which $f(x) = 0$. This means that the roots of the function are the x-values at the points where $y = f(x)$ crosses the x-axis.

Now try this

The diagram shows a sketch of the curve with equation $y = x^2 - 4x - 3$

(a) Write down the coordinates of the point A, where the curve crosses the y-axis. **(1 mark)**

(b) By completing the square for $x^2 - 4x - 3$ find the coordinates of the turning point B. **(3 marks)**

Worked solution video

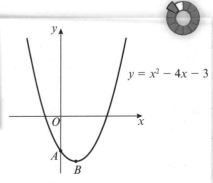

$y = x^2 - 4x - 3$

A

B

Sketching graphs

A **sketch** of a graph shows all the major features. These usually include turning points and places where the graph crosses the axes.

Factors and roots

If a function is **factorised** you can work out the points where its graph crosses the x-axis. These are called the **roots** of the function. You can find them by working out the x-values that make each factor equal to **zero**.

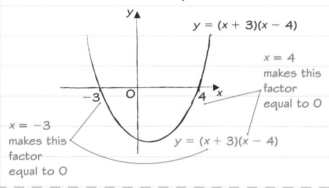

$y = (x + 3)(x - 4)$

$x = 4$ makes this factor equal to 0

$x = -3$ makes this factor equal to 0

$y = (x + 3)(x - 4)$

Graph sketching checklist

- ☑ Your sketch should be **clear** and **neat**.
- ☑ Use a **ruler** for any **straight** lines.
- ☑ Label your **axes** and the **origin**.
- ☑ Label your graph with its **equation**.
- ☑ Label any turning points.

There is more about finding turning points of quadratic graphs on page 43.

- ☑ Label any points where the graph crosses the axes.

The question will usually state exactly what points you need to show.

Sketching cubics – two to watch

 1 If one factor is x then the curve will pass through the origin.

2 If one factor is **squared** then the curve will **just touch** the x-axis at the corresponding point.

The factor $(x + 3)$ is **squared** (or **repeated**). This means that the curve **just touches** the x-axis at the point that makes this factor equal to zero. Be careful with your x-values. The factor $(x + 3)$ is zero when $x = -3$ and the factor $(x - 2)$ is zero when $x = 2$.

Don't forget to work out the point where the curve intercepts the y-axis. You do this by setting $x = 0$ in the function:
$y = (0 - 2)(0 + 3)^2 = -2 \times 9 = -18$

$f(x) = (x - 2)(x + 3)^2$
Sketch the graph of $y = f(x)$, showing the coordinates of any intercepts with the coordinate axes.   **(3 marks)**

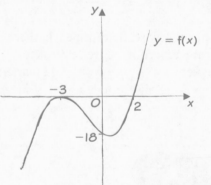

$y = f(x)$

Now try this

 Sketch graphs of the following equations, showing the coordinates of any intercepts with the coordinate axes.

(a) $y = x^2 - 7x + 12$ **(2 marks)**

(b) $y = x(x^2 + 4x - 12)$ **(3 marks)**

Iteration

You can find exact answers to quadratic equations by completing the square, or using the quadratic formula. It can be trickier to find exact answers to more complicated equations, like those involving cubes or square roots. You can use **iteration** to find numerical answers to a given degree of accuracy.

Iteration formula

Here is a cubic function: $f(x) = x^3 - 5x + 3$
The **roots** of this function occur when $f(x) = 0$.
These are the points where the graph of $y = f(x)$ crosses the x-axis.

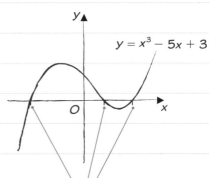

$y = x^3 - 5x + 3$

You can use an **iteration formula** to find the roots. Here is an iteration formula for this function:

$$x_{n+1} = \sqrt[3]{5x_n - 3}$$

You input a value for x here.

The **output** from the formula becomes your **next** input value.

This function has 3 different real roots. The iteration formula might produce different roots depending on the starting value, x_0. You will usually be told what value of x_0 to use.

If you have to use an iteration formula to find a root or solve an equation, you will usually be given it in the exam.

Part (a) shows you how you can find an iteration formula for a particular equation.

Worked example

The function $f(x) = x^3 - x - 2$ has a real root between 1 and 2.
(a) Show that when $f(x) = 0$, $x = \sqrt[3]{x + 2}$. **(2 marks)**

$x^3 - x - 2 = 0$ $(+2)$
$x^3 - x = 2$ $(+x)$
$x^3 = x + 2$ $(\sqrt[3]{})$
$x = \sqrt[3]{x + 2}$

(b) Use the iterative formula $x_{n+1} = \sqrt[3]{x_n + 2}$ with $x_0 = 1$ to find the real root of $f(x)$ correct to 4 decimal places. **(3 marks)**

$x_0 = 1$
$x_1 = \sqrt[3]{x_0 + 2} = 1.442\,249\,57... = 1.4422$ (4 d.p.)
$x_2 = \sqrt[3]{x_1 + 2} = 1.509\,897\,449... = 1.5099$ (4 d.p.)
$x_3 = \sqrt[3]{x_2 + 2} = 1.519\,724\,305... = 1.5197$ (4 d.p.)
$x_4 = \sqrt[3]{x_3 + 2} = 1.521\,141\,269... = 1.5211$ (4 d.p.)
$x_5 = \sqrt[3]{x_4 + 2} = 1.521\,345\,368... = 1.5213$ (4 d.p.)
$x_6 = \sqrt[3]{x_5 + 2} = 1.521\,374\,761... = \underline{1.5214}$ (4 d.p.)
$x_7 = \sqrt[3]{x_6 + 2} = 1.521\,3789\,95... = \underline{1.5214}$ (4 d.p.)

Using a calculator

The Ans button enters the answer to the previous calculation. This is exactly what you need for speedy iterations! In the worked example on the left you could use these keys for each new line of working:

You have to find the solution correct to 4 decimal places. Start with the value of x_0 given in the question. Work out x_1, x_2, x_3 and so on. Write down all the digits shown on your calculator display then round each value to 4 d.p. When you reach **two values** that both round to the same number you can stop. That number is your answer correct to 4 d.p.

Now try this

The function $f(x)$ is defined by $f(x) = x + \sqrt{x} - 5$.

(a) Show that when $f(x) = 0$, $x = 5 - \sqrt{x}$ **(1 mark)**

(b) Use the iterative formula $x_{n+1} = 5 - \sqrt{x_n}$ with $x_0 = 1$ to find the real root of $f(x)$ correct to 3 decimal places. **(3 marks)**

Rearranging formulae

Most formulae have one letter on its own on one side of the formula. This letter is called the **subject** of the formula.

$$E = mc^2 \qquad E \text{ is the subject of the formula.}$$

Changing the subject of a formula is like solving an equation. You have to do the same thing to both sides of the formula until you have the new letter on its own on one side.

$$E = mc^2 \qquad (\div m)$$

$$\frac{E}{m} = c^2 \qquad (\sqrt{\ })$$

The inverse operation to $\boxed{x^2}$ is $\boxed{\sqrt{\ }}$. You need to square root **everything** on both sides of the formula.

$$\sqrt{\frac{E}{m}} = c \qquad c \text{ is now the subject of the formula.}$$

Harder formulae

If the letter you need **appears twice** in the formula you need to **factorise**.

| **Group** all the terms with that letter on one side of the formula and all the other terms on the other side. | → | **Factorise** so the letter only appears once. | → | **Divide** by everything in the bracket to get the letter on its own. |

For a reminder about factorising have a look at page 18.

Worked example

$$N = \frac{3h + 20}{100}$$

Rearrange the formula to make h the subject.
(2 marks)

$$N = \frac{3h + 20}{100} \qquad (\times 100)$$

$$100N = 3h + 20 \qquad (-20)$$

$$100N - 20 = 3h \qquad (\div 3)$$

$$\frac{100N - 20}{3} = h$$

$$h = \frac{100N - 20}{3}$$

Your final answer should look like $h = \ldots$

Worked example

Make Q the subject of the formula $P = \dfrac{Q}{Q - 100}$

(3 marks)

$$P = \frac{Q}{Q - 100} \qquad [\times (Q - 100)]$$

$$P(Q - 100) = Q \qquad \text{(multiply out brackets)}$$

$$PQ - 100P = Q \qquad (+ 100P)$$

$$PQ = Q + 100P \qquad (- Q)$$

$$PQ - Q = 100P \qquad \text{(factorise)}$$

$$Q(P - 1) = 100P \qquad [\div (P - 1)]$$

$$Q = \frac{100P}{P - 1}$$

Your final answer should look like $Q = \ldots$
You need to factorise to get Q on its own.

Now try this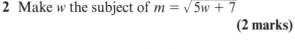

 1 Rearrange this formula to make t the subject:
$4p = 3t - 1$ **(2 marks)**

2 Make w the subject of $m = \sqrt{5w + 7}$
(2 marks)

 3 Rearrange this formula to make P the subject

$$Q = \sqrt{\frac{100 - 5P}{P}}$$

(4 marks)

Worked solution video

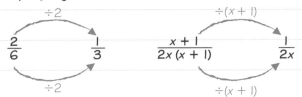

Algebraic fractions

Simplifying an algebraic fraction is just like simplifying a normal fraction.

$$\frac{2}{6} \xrightarrow{\div 2} \frac{1}{3}$$ ($\div 2$)

$$\frac{x+1}{2x(x+1)} \xrightarrow{\div(x+1)} \frac{1}{2x}$$ ($\div(x+1)$)

You can divide the top and bottom by a number, a term, or a whole expression.

Golden rule

If the top or the bottom of the fraction has **more than** one term, you will need to factorise before simplifying.

$$\frac{p^2 + 3p}{4p} = \frac{p(p+3)}{4p} = \frac{p+3}{4}$$

Two terms on top so factorise the top, then divide the top and bottom by p.

Operations on algebraic fractions

1 To **add** or **subtract** algebraic fractions with different denominators:

1. Find a common denominator.
2. Add or subtract the numerators.
3. Simplify if possible.

$$\frac{1}{x+4} + \frac{2}{x-4} = \frac{x-4}{(x+4)(x-4)} + \frac{2(x+4)}{(x+4)(x-4)}$$

$$= \frac{x-4+2x+8}{(x+4)(x-4)} = \frac{3x+4}{(x+4)(x-4)}$$

The smallest common denominator isn't always the product of the two denominators.

You can use a common denominator of $4x$

to simplify this expression:

$$\frac{x+1}{2x} + \frac{3-2x}{4x}$$

2 To **multiply** fractions:

1. Multiply the numerators **and** multiply the denominators.
2. Simplify if possible.

$$\frac{x}{2} \times \frac{4}{x-1} = \frac{\overset{2}{\cancel{4}}x}{\underset{1}{\cancel{2}}(x-1)} = \frac{2x}{x-1}$$

Don't expand brackets if you don't have to. It's much easier to simplify your fraction with the brackets in place.

3 To **divide** fractions:

1. Change the second fraction to its reciprocal.
2. Change ÷ to ×
3. Multiply the fractions and simplify.

$$\frac{x^2}{3} \div \frac{x}{6} = \frac{x^2}{3} \times \frac{6}{x} = \frac{\overset{2}{\cancel{6}}x^{\cancel{2}}}{\underset{1}{\cancel{3}}x} = 2x$$

To find the reciprocal of a fraction you turn it upside down.

Worked example

Simplify $\dfrac{3x^2 - 8x - 3}{x^2 - 9}$ **(3 marks)**

$$\frac{3x^2 - 8x - 3}{x^2 - 9} = \frac{(3x+1)(x-3)}{(x+3)(x-3)}$$

$$= \frac{3x+1}{x+3}$$

You need to factorise the top and the bottom of the fraction before you can simplify. Remember that $a^2 - b^2 = (a+b)(a-b)$

Now try this

1 Simplify fully $\dfrac{a+1}{3a} + \dfrac{7}{6a}$ **(2 marks)**

2 Solve $\dfrac{1}{10x} - \dfrac{1}{15x} = 4$ **(3 marks)**

3 Simplify fully

(a) $\dfrac{x^2 + 4x - 12}{x^2 - 25} \div \dfrac{x+6}{x^2 - 5x}$ **(4 marks)**

(b) $\dfrac{3m^2 - 108}{9m^3 + 54m^2}$ **(3 marks)**

Worked solution video

Quadratics and fractions

You need to remove fractions before you can solve an equation.

For a reminder about solving linear equations with fractions have a look at page 20.

To remove fractions from an equation multiply everything by the lowest common multiple of the denominators.

$$\frac{x}{2x-3} + \frac{4}{x+1} = 1$$

(2x − 3) and (x + 1) don't have any common factors. Multiply everything by (2x − 3)(x + 1)

$$\frac{x\cancel{(2x-3)}(x+1)}{\cancel{2x-3}} + \frac{4(2x-3)\cancel{(x+1)}}{\cancel{x+1}} = (2x-3)(x+1)$$

Don't expand brackets until you have simplified the fractions.

$$x(x+1) + 4(2x-3) = (2x-3)(x+1)$$
$$x^2 + x + 8x - 12 = 2x^2 - 3x + 2x - 3$$
$$0 = x^2 - 10x + 9$$
$$= (x-9)(x-1)$$

Multiply out brackets and collect like terms.

The solutions are $x = 9$ and $x = 1$

Worked example

Solve $\dfrac{2+3x}{5x+9} = \dfrac{2}{x-1}$ **(4 marks)**

$$\left(\frac{2+3x}{5x+9}\right)(x-1)(5x+9) = \left(\frac{2}{x-1}\right)(x-1)(5x+9)$$

$$\frac{(2+3x)(x-1)\cancel{(5x+9)}}{\cancel{5x+9}} = \frac{2\cancel{(x-1)}(5x+9)}{\cancel{x-1}}$$

$$(2+3x)(x-1) = 2(5x+9)$$
$$3x^2 - 3x + 2x - 2 = 10x + 18$$
$$3x^2 - 11x - 20 = 0$$
$$(3x+4)(x-5) = 0$$

$$3x + 4 = 0 \qquad x - 5 = 0$$
$$x = -\frac{4}{3} \qquad x = 5$$

Multiply everything by $(x - 1)(5x + 9)$ to remove the fractions.

If you are confident working with algebraic fractions you can jump straight to this step.

Set the two factors equal to 0 to find the two solutions.

Quadratic equations checklist

Remove any fractions by multiplying everything by the lowest common multiple of the denominators. ✓

Multiply out any brackets and collect like terms. ✓

Rewrite in the form $ax^2 + bx + c = 0$ ✓

Factorise the left-hand side to solve the quadratic equation. ✓

Now try this

Multiply everything by $5(3x - 1)(2x + 1)$

1 Solve $\dfrac{5}{x} + \dfrac{2}{x+2} = 3$ **(4 marks)**

Worked solution video

2 Solve $\dfrac{2}{3x-1} - \dfrac{3}{2x+1} = \dfrac{2}{5}$ **(4 marks)**

3 Solve $\dfrac{1}{2x+3} - \dfrac{1}{x} = \dfrac{1}{20}$ **(4 marks)**

Surds 2

You might need to expand brackets involving surds. You can use the grid method, or **FOIL** to expand the brackets (have a look at page 17). Here are two **golden rules** to remember when working with surds.

 You can use the rule $\sqrt{ab} = \sqrt{a} \times \sqrt{b}$ in **both directions**:

$\sqrt{20} \times \sqrt{5} = \sqrt{100} = 10$

$\sqrt{2a} \times \sqrt{18a} = \sqrt{36a^2} = \sqrt{36} \times \sqrt{a^2} = 6a$

 Whole number parts and surd parts stay **separate**:

If $24 + 6\sqrt{7} = a + b\sqrt{7}$ then you can **compare** the two expressions to get $a = 24$ and $b = 6$

Worked example

Show that $(3 + \sqrt{8})(4 + \sqrt{8}) = 20 + 14\sqrt{2}$

Show each stage of your working clearly. **(2 marks)**

$(3 + \sqrt{8})(4 + \sqrt{8}) = 12 + 3\sqrt{8} + 4\sqrt{8} + (\sqrt{8})^2$

$= 20 + 7\sqrt{8}$

$= 20 + 7 \times 2\sqrt{2}$

$= 20 + 14\sqrt{2}$

When the question says 'Show that …' you should start from the **left-hand side**, then simplify and rearrange until your expression matches the **right-hand side**.

Make sure you **simplify** any surds in your expression:

$\sqrt{8} = \sqrt{4 \times 2}$

$= \sqrt{4} \times \sqrt{2}$

$= 2\sqrt{2}$

There's more about this on page 12.

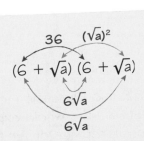

$$(6 + \sqrt{a})(6 + \sqrt{a})$$

Simplify the expression, remembering that $(\sqrt{a})^2 = a$, then compare your answer with the right-hand side given in the question. $12\sqrt{a} = 12\sqrt{2}$ so $a = 2$, and $36 + a = b$ so $b = 38$

Worked example

Given that a and b are positive integers such that $(6 + \sqrt{a})^2 = b + 12\sqrt{2}$

find the value of a and the value of b. **(3 marks)**

$(6 + \sqrt{a^2}) = (6 + \sqrt{a})(6 + \sqrt{a})$

$= 36 + 6\sqrt{a} + 6\sqrt{a} + (\sqrt{a})^2$

$= 36 + a + 12\sqrt{a}$

$= b + 12\sqrt{2}$

$a = 2$ and $b = 38$

Now try this

1 Show that $= (3 - \sqrt{12})^2 = 21 - 12\sqrt{3}$
Show each stage of your working clearly. **(3 marks)**

2 $(5 + \sqrt{x})(3 + \sqrt{x}) = 18 + k\sqrt{x}$
where x is a prime number and k is a positive integer.
Find the value of x and the value of k. **(3 marks)**

Functions

You need to be able to work with **function notation** confidently in your exam.

f is the **name** of the function. You can use any letter, but f and g are the most common.

$$f(x) = \sqrt{x - 2}$$

x is the **input**. You say 'f of x'. You can also write $f : x \rightarrow \sqrt{x - 2}$ and say 'f maps x onto $\sqrt{x - 2}$'

This tells you what the function does to x.

The function f is such that $f(x) = 3x + 10$

(a) Find $f(-2)$ **(1 mark)**

$f(-2) = 3 \times (-2) + 10$
$= -6 + 10 = 4$

(b) Solve $f(a) = 31$ **(2 marks)**

$3a + 10 = 31 \qquad (-10)$
$\qquad 3a = 21 \qquad (\div 3)$
$\qquad a = 7$

Composite functions

If you apply two functions one after the other, you can write a **single function** which has the same effect as the two combined functions. This is called a **composite function**.

The function gf has the same effect as applying function f **then** applying function g.

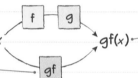

The **order** is important. The function being applied **first** goes **closest** to the x.

Order is important in composite functions. You can think of fg(x) as f[g(x)]

You work out g(x) **first**, then you use this answer as your **input** for f(x).

Note that in this case gf(10) would give you a **different answer**:

$f(10) = 10^2 = 100$
$g(100) = 100 - 3 = 97$
So gf(10) = 97

$f(x) = x^2$
$g(x) = x - 3$
Find fg(10) **(2 marks)**

$g(10) = 10 - 3 = 7$
$f(7) = 7^2 = 49$
So fg(10) = 49

$f(x) = 3x + 2$
$g(x) = \dfrac{x}{x + 2}$

Find fg(x)

Give your answer as a single algebraic fraction expressed as simply as possible. **(3 marks)**

$fg(x) = f[g(x)] = f\left[\dfrac{x}{x + 2}\right]$

$= 3\left[\dfrac{x}{x + 2}\right] + 2$

$= \dfrac{3x}{x + 2} + 2$

$= \dfrac{3x}{x + 2} + \dfrac{2(x + 2)}{x + 2}$

$= \dfrac{3x + 2x + 4}{x + 2} = \dfrac{5x + 4}{x + 2}$

To find an algebraic expression for fg(x) you need to substitute the **whole expression** for g(x) for each instance of x in the expression for f(x).

1 The function f is defined as
$$f(x) = \frac{x + 3}{x}$$
(a) Find $f(1)$ **(1 mark)**
(b) Given that $f(b) = 3$, find the value of b. **(2 marks)**

2 The functions f and g are such that
$f(x) = 2x - 2$ and $g(x) = x^2 - 4$
(a) Find gf(x)
Give your answer as simply as possible. **(2 marks)**
(b) Solve gf(x) = 0 **(3 marks)**

Inverse functions

For a function f, the **inverse** of f is the function that **undoes** f. You write the inverse as f^{-1}. If you apply f then f^{-1}, you will end up back where you started.

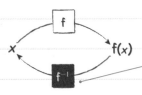

If you apply f then f^{-1}, you have applied the **composite function** $f^{-1}f$. The output of $f^{-1}f$ is the **same** as the input. You can write:
$$f^{-1}f(x) = ff^{-1}(x) = x$$

Finding the inverse

To find the inverse of a function given in the form $f(x) = \ldots$ you need to:

 Write the function in the form $y = \ldots$

 Rearrange to make x the subject

 Swap any y's for x's and rewrite as $f^{-1}(x) = \ldots$

For a reminder on changing the subject of a formula, have a look at page 46.

$$g(x) = \frac{3}{x + 4}$$

Express the inverse function g^{-1} in the form $g^{-1}(x) = \ldots$ **(3 marks)**

$$y = \frac{3}{x + 4} \qquad (\times (x + 4))$$
$$y(x + 4) = 3 \qquad (\div y)$$
$$x + 4 = \frac{3}{y} \qquad (- 4)$$
$$x = \frac{3}{y} - 4 \qquad \text{(Swap } y\text{'s for } x\text{'s)}$$
$$g^{-1}(x) = \frac{3}{x} - 4$$

Start by multiplying both sides by $(x + 4)$. You want x on its own on the left-hand side so don't expand the bracket. Once you've rearranged to make x the subject, swap any y's for x's and and write your answer as $g^{-1}(x) = \ldots$

Worked example

The function f is such that $f(x) = 3x + 5$
Express the inverse function f^{-1} in the form $f^{-1}(x) = \ldots$ **(2 marks)**

$$y = 3x + 5 \qquad (- 5)$$
$$y - 5 = 3x \qquad (\div 3)$$
$$\frac{y - 5}{3} = x \qquad \text{(Swap } y\text{'s for } x\text{'s)}$$
$$f^{-1}(x) = \frac{x - 5}{3}$$

You can also use a flow chart to find an inverse.
Here is a flow chart for $f(x)$

$$x \rightarrow \boxed{\times 3} \rightarrow \boxed{+5} \rightarrow f(x)$$

You work **backwards** through the flow chart to find $f^{-1}(x)$

$$f^{-1}(x) \leftarrow \boxed{\div 3} \leftarrow \boxed{-5} \leftarrow x$$

You subtract 5, **then** divide by 3. Written using algebra this is
$$f^{-1}(x) = \frac{x - 5}{3}$$

1 The function f is such that $f(x) = 2x - 1$
   (a) Express the inverse function f^{-1} in the form $f^{-1}(x) = \ldots$ **(2 marks)**
   (b) Without doing any further working, write down the value of $f^{-1}f(7)$ **(1 mark)**

2 The function g is defined as
   $$g(x) = \frac{x + 6}{x}$$ Express the inverse function g^{-1} in the form $g^{-1}(x) = \ldots$ **(3 marks)**

Group all the terms involving x on one side, then factorise to get x on its own.

Algebraic proof

You can use algebra to **prove** facts about numbers.

Using algebra helps you to prove that something is true for **every** number.

In a proof question the working **is** the answer.

If the question says '**Show that** ...' or '**Prove that** ...', you need to write down every stage of your working. If you don't, you won't get all the marks.

Golden rule

If you need to prove something about numbers then you always use algebra.

Algebraic proof toolkit

Use n to represent any whole number.

Number fact	Written using algebra
Even number	$2n$
Odd number	$2n + 1$ or $2n - 1$
Multiple of 3	$3n$
Consecutive numbers	$n, n + 1, n + 2, ...$
Consecutive even numbers	$2n, 2n + 2, 2n + 4, ...$
Consecutive odd numbers	$2n + 1, 2n + 3, 2n + 5, ...$
Consecutive square numbers	$n^2, (n + 1)^2, (n + 2)^2, ...$

Problem solved!

To **prove** the statement you need to show that it is true for **any** three consecutive integers. You can do this using algebra.

1. Write the first integer as n and the next two integers as $n + 1$ and $n + 2$.
2. Write an expression for the sum of your three integers.
3. Simplify and factorise the expression.
4. Explain why the final expression is divisible by 3.

You will need to use problem-solving skills throughout your exam – **be prepared!**

Worked example

Prove that the sum of three consecutive integers is always divisible by 3. **(3 marks)**

$n, n + 1$ and $n + 2$ represent any three consecutive integers.

$n + (n + 1) + (n + 2) = 3n + 3$
$\qquad\qquad\qquad\qquad = 3(n + 1)$

$n + 1$ is an integer, so $3(n + 1)$ is divisible by 3.

Identities

The symbol $\equiv$ represents an **identity**. This is something which is always true. If you have to show that an identity is true in your exam

- start with the left-hand side
- manipulate it until it matches the right-hand side.

Worked example

Show that $(a + b)(a - b) \equiv a^2 - b^2$ **(2 marks)**

$(a + b)(a - b) = a^2 - ab + ab - b^2$
$\qquad\qquad\qquad = a^2 - b^2$

An identity is not an equation – don't try to solve it using the balance method.

Now try this

1 The nth term for the sequence of triangular numbers is
$\frac{1}{2}n(n + 1)$
Prove that the sum of two consecutive triangular numbers is a square number. **(3 marks)**

Worked solution video

2 Two integers have a difference of 4. The difference between the squares of the two integers is eight times the mean of the integers.
For example, $14 - 10 = 4$
$14^2 - 10^2 = 196 - 100 = 96$

The mean of 10 and 14 is $\frac{10 + 14}{2} = 12$
and $8 \times 12 = 96$

Prove this result algebraically. **(4 marks)**

Write the two integers as n and $(n + 4)$.

Exponential graphs

A function of the form f(x) = ka^x is called an **exponential function**. You need to be able to recognise and sketch the graphs of exponential functions. The shape of an exponential graph depends on whether x is positive or negative.

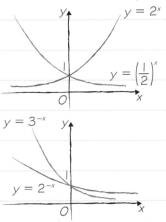

Exponential growth

Exponential graphs appear in equations representing growth or decay. If there are 500 bacteria in a Petri dish, and the number doubles every hour, then after h hours there will be $N = 500 \times 2^h$ bacteria in the Petri dish. The graph of N against h will be an exponential graph.

There is more about exponential growth and decay on page 64.

Worked example

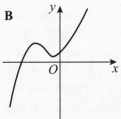

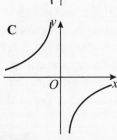
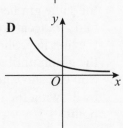

Write down the letter of the graph which could have the equation

(a) $y = 0.5^{-x}$ ....D....

(b) $y = \dfrac{2}{x}$ ....A....

(c) $y = x^3 + 3x^2 + 2x + 1$ ....B.... **(3 marks)**

Finding missing values

Some questions ask you to find the missing value in an exponential equation.

This graph has equation $y = ka^x$. You can substitute the values of x and y that you are given to get two equations.

$$7 = ka^1 \qquad ①$$
$$175 = ka^3 \qquad ②$$
$$ka^3 \div ka^1 = 175 \div 7 \qquad ② \div ①$$
$$a^2 = 25$$
$$a = 5$$

Substitute a = 5 into ①:
$$7 = 5k$$
$$k = \frac{7}{5}$$

Check it!
Substitute in ②:
$$ka^3 = \frac{7}{5}(5)^3$$
$$= 7 \times 25$$
$$= 175 ✓$$

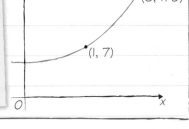

Now try this

A small fishing lake is stocked with trout. The number of trout in the lake T weeks after the stocking date is modelled by the formula
$$N = 80 \times 0.8^T$$

(a) How many trout were put into the lake on the stocking date? **(1 marks)**

(b) Estimate the number of trout remaining after 2 weeks. **(2 marks)**

(c) Draw a graph to show the number of trout in the lake for the first 15 weeks. **(3 marks)**

(d) Use your graph to estimate the length of time it will take for the number of trout in the lake to reduce by 75%. **(2 marks)**

Gradients of curves

You can estimate the gradient of a curve at a given point by drawing a **tangent** to the curve at that point. This distance–time graph shows a runner accelerating at the start of a race.

On a distance–time graph the gradient tells you the speed. There is more about this on page 30.

Distance (m)

Pay attention to the scale when you calculate the gradient.

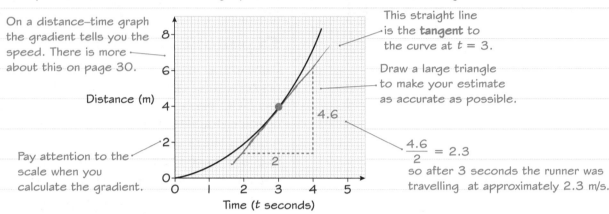

This straight line is the **tangent** to the curve at $t = 3$.

Draw a large triangle to make your estimate as accurate as possible.

$\frac{4.6}{2} = 2.3$

so after 3 seconds the runner was travelling at approximately 2.3 m/s.

Time (t seconds)

Worked example

This graph shows the voltage across a phone battery as it charges from empty.

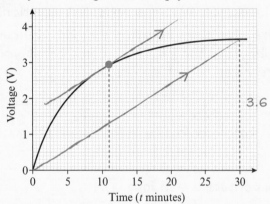

Voltage (V)

Time (t minutes)

(a) Work out the average rate of increase of voltage between $t = 0$ and $t = 30$. **(2 marks)**

$\frac{3.6}{30} = 0.12$ V/min

Amir wants to stop charging his phone when the rate of increase of voltage drops to this average level.

(b) After how long should Amir stop charging his phone? You must show how you got your answer. **(2 marks)**

11 minutes

Problem solved!

(a) To work out the average rate of change between $t = 0$ and $t = 30$ draw a straight line between these points on the graph and find its gradient. This is the same as working out

$$\frac{\text{change in voltage}}{\text{change in time}}$$

Look at the axis labels to work out the correct units.

(b) You need to find the point **on the curve** with the same gradient. That means you need to find a **tangent** to the curve that is parallel to the first line. Slide a transparent ruler across the graph until it just touches the curve. You can show how you got your answer by drawing the tangent to the curve at this point.

You will need to use problem-solving skills throughout your exam – **be prepared!**

Now try this

A container is filling with water. This graph shows the depth of water in the container.

(a) Work out the average rate of increase of depth between $t = 0$ and $t = 40$. **(2 marks)**

(b) Use the graph to estimate the rate of increase of depth of water at $t = 10$. **(2 marks)**

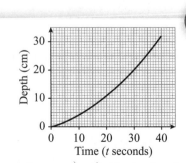

Depth (cm)

Time (t seconds)

Velocity–time graphs

Velocity means **speed** in a certain direction. A velocity–time graph is sometimes called a **speed–time graph**. You need to be able to interpret a velocity–time graph. Use the two golden rules on the right to answer questions involving speed, time and acceleration.

This velocity–time graph shows the motion of a remote-controlled car.

The car was travelling at a **constant speed** of 2 m/s for this section of the journey.

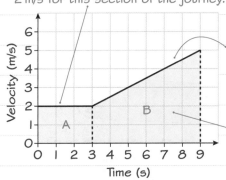

During this part of the journey the car was accelerating at a rate of:

$$\frac{5-2}{9-3} = \frac{3}{6} = 0.5 \text{ m/s}^2$$

Treat this shape as a **trapezium** to work out the area.

Area A = 2 × 3 = 6
Area B = ½(2 + 5) × 6 = 21
Total **distance travelled**
= 6 + 21 = 27 m

Worked example

A boat travels at a constant speed of 15 m/s for 40 seconds. It then decelerates to a speed of 8 m/s in 24 seconds.

Assuming the deceleration is constant, calculate the total distance travelled by the boat. **(4 marks)**

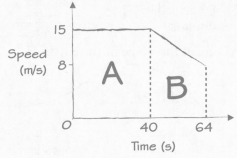

Area A = 15 × 40 = 600
Area B = ½(15 + 8) × 24 = 276
Total distance travelled = 600 + 276
= 876 m

Problem solved!

The easiest way to solve this problem is to sketch a velocity–time graph of the journey. The graph doesn't have to be to scale, but make sure you label the axes and mark any important points. **Constant** deceleration means that this section of the velocity–time graph will be a straight line. To calculate the total distance travelled you need to work out the total area underneath the graph. Divide it into sections, work out the area of each section, and then add them together.

You will need to use problem-solving skills throughout your exam – **be prepared!**

Be careful when you are calculating areas. You need to read values off the **scale**. Don't just count grid squares. The units of speed are m/s and the units of time are seconds, so the distance travelled will be in metres.

Now try this

Ashley gets on her bike and accelerates constantly to a speed of 4.5 m/s in 20 seconds. She remains at this constant speed for a further 40 seconds. She then decelerates constantly to rest in 8 seconds.
(a) Sketch a velocity–time graph of Ashley's journey. **(3 marks)**
(b) Calculate the total distance Ashley travelled. **(2 marks)**

Areas under curves

You need to be able to estimate the area under a curve. You can do this by drawing trapeziums in **equal intervals** underneath the graph.

The diagram shows the graph of $y = x^3 - 6x^2 + 8x + 6$. You can estimate the area under this curve between $x = 1$ and $x = 4$ by drawing 3 trapeziums of equal width.

Area of A $= \frac{1}{2}(9 + 6) \times 1 = 7.5$
Area of B $= \frac{1}{2}(6 + 3) \times 1 = 4.5$
Area of C $= \frac{1}{2}(3 + 6) \times 1 = 4.5$

So the estimate for the area is
$7.5 + 4.5 + 4.5 = 16.5$

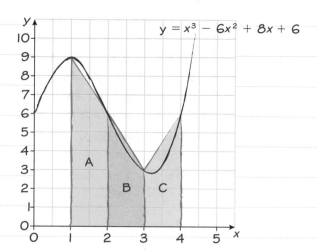

Worked example

This velocity–time graph shows part of a rollercoaster ride.

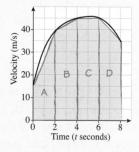

Everything in red is part of the answer.

(a) Use four equal intervals on the graph to estimate the total distance travelled by the rollercoaster between $t = 0$ and $t = 8$.

(4 marks)

$\frac{1}{2}(15 + 40) \times 2 + \frac{1}{2}(40 + 45) \times 2$
$+ \frac{1}{2}(45 + 45) \times 2 + \frac{1}{2}(45 + 35) \times 2$
$= 55 + 85 + 90 + 80 = 310 \, \text{m}$

(b) Is your answer to part (a) an overestimate or an underestimate. Justify your answer.

(1 mark)

Underestimate. Every trapezium is completely underneath the curve.

Problem solved!

(a) The area under a velocity–time graph tells you the distance travelled. Divide the area into four equal intervals and draw trapeziums. If you label your trapeziums it will make your working clearer. The formula for the area of a trapezium is

Area $= \frac{1}{2}(a + b)h$

It's easiest to think of the **width** of the trapezium as h and a and b as the heights of the endpoints.

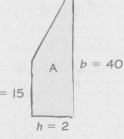

(b) Look at the shape of the curve.

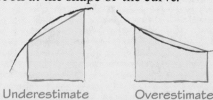

Underestimate   Overestimate

You will need to use problem-solving skills throughout your exam – **be prepared!**

Now try this

The diagram shows the graph of a cubic function, $y = \text{f}(x)$. Use three equal intervals to estimate the area under the curve between $x = 2$ and $x = 8$.

(4 marks)

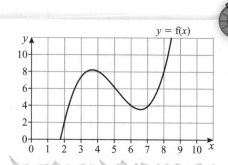

Problem-solving practice 1

Throughout your Higher GCSE exam you will need to **problem-solve**, **reason**, **interpret** and **communicate** mathematically. If you come across a tricky or unfamiliar question in your exam you can try some of these strategies:

- ✓ Sketch a diagram to see what is going on.
- ✓ Try the problem with smaller or easier numbers.
- ✓ Plan your strategy before you start.
- ✓ Write down any formulae you might be able to use.
- ✓ Use x or n to represent an unknown value.

---

1   Here are the first four terms of an arithmetic sequence.

9, 15, 21, 27, ...

Worked solution video

Explain why the number 65 cannot be a term of this sequence.   **(3 marks)**

Arithmetic sequences page 22

Work out the nth term of the sequence. You could set it equal to 65 and solve an equation to show that n is not an integer, or you could work out consecutive terms of the sequence on either side of 65.

TOP TIP

You can use any successful strategy to answer a question, as long as you show your method.

---

2   A train travels a distance of y km in 3 minutes. Show that the average speed of the train is $20y$ km/h.   **(3 marks)**

Formulae page 21
Speed page 65

Watch out! The time is given in minutes, but you need to find an expression for the speed in km/h. Divide by 60 then simplify your fraction. Because this is a "Show that..." question you need to write down your conversion **clearly** and give the **units**.

TOP TIP

Try changing y into a number and calculating the average speed. It will give you an idea of how the formula works.

---

3   A straight line has gradient 8 and passes through the point (5, 10).

(a) Find the equation of the line.   **(2 marks)**

(b) Determine whether the point (10, 50) lies on the line.   **(1 mark)**

Straight-line graphs 2 page 26

You could tackle this question using algebra, or by drawing a diagram to scale. Whichever method you use, make sure you write a conclusion.

TOP TIP

If you have to comment on **reliability** then think about the method you used. If you used a diagram your answer might be less reliable than if you used algebra.

Problem-solving practice 2

4 A music download website sells singles and albums.

Worked solution video

The music download website makes 1000 sales in one hour.

Singles cost 50p to download.

Albums cost £4.50 to download.

The income for this hour is £1500.

How many singles does the website sell in this hour? **(4 marks)**

Simultaneous equations 1 page 34

You need to use the information in the question to write two simultaneous equations. Choose one letter to represent the number of albums and a different letter to represent the number of singles.

TOP TIP

If you are comparing quantities, make sure that the **units** are consistent. In this question you need to work either in pounds or pence – not both!

---

5 The function f is defined as $f(x) = 2x + 15$

a is a positive integer.

$f(a) = b$

$f(b) = a^2$

Work out the value of a. **(4 marks)**

Functions page 50

Write an expression for b in terms of a. Then use this expression as the input for the function. Set the output equal to a^2 and solve the equation to find a.

TOP TIP

If the **input** to a function is an **expression**, use brackets when you substitute.

---

6 C is the circle with equation $x^2 + y^2 = 20$

L is the line with equation $y = 2x - 10$

Prove that **L** is a tangent to **C**. **(6 marks)**

Equation of a circle page 36
Simultaneous equations 2 page 35

The easiest way to answer this question is to find the point where the line intersects the circle. A straight line can intersect a circle at 0, 1 or 2 points. If the line intersects the circle at exactly 1 point then it is a tangent to the circle.

TOP TIP

If you have to answer a question about equations of graphs and there is no diagram, it is a good idea to draw a sketch.

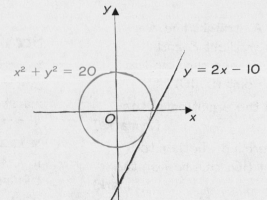

$x^2 + y^2 = 20$ $y = 2x - 10$

Calculator skills 2

You need to be able to work out basic percentages quickly using your calculator.

Calculating with fractions

You can enter fractions on your calculator using the ▢ key and the arrows.
For example, to work out $\frac{3}{8}$ of 52:

$$\frac{3}{8} \times 52$$

$$\frac{39}{2}$$

If you want to convert an answer on your calculator display from a fraction to a decimal you can use the S⇔D key.

Quick conversions

You can convert between fractions, decimals and percentages quickly using a calculator:

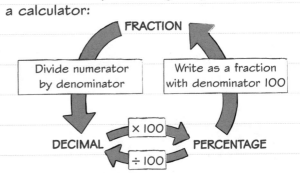

FRACTION

Divide numerator by denominator | Write as a fraction with denominator 100

DECIMAL   × 100 / ÷ 100   PERCENTAGE

You can solve lots of percentage problems by working out what 1% represents. To find a percentage of an amount:

Divide the percentage by 100

↓

Multiply by the amount

Worked example

A company gives 3.5% of its profits to charity. In 2011 the company made profits of £470 000. How much money did the company give to charity in 2011? **(2 marks)**

$3.5 \div 100 = 0.035$

$0.035 \times 470\,000 = 16\,450$

The company gave £16 450 to charity.

Worked example

In a year group of 85 students, 62 buy their lunch at school. Express 62 as a percentage of 85. Give your answer correct to 1 decimal place. **(2 marks)**

$62 \div 85 = 0.72941...$

$0.72941... \times 100 = 72.941...$

72.9% of students buy their lunch in school.

To write one quantity as a percentage of another:

Divide the first quantity by the second quantity

↓

Multiply your answer by 100

Always write down at least five digits from your calculator display before rounding your answer.

Now try this

1 Last year a university had 226 graduates. 195 of them found jobs immediately. Express 195 as a percentage of 226. Give your answer correct to 1 decimal place. **(2 marks)**

2 Aisha earns £2230 per month and spends 25% of it on rent. Joshua earns £1800 per month and spends 30% on rent. Who spends the greater amount on rent? **(3 marks)**

You need to show your working clearly to demonstrate your strategy.

Ratio

Ratios are used to compare quantities.
You can find equivalent ratios by multiplying or dividing by the same number.

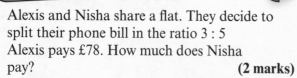

$5 : 9$

×2 ↓ ↓ ×2

$10 : 18$

÷10 ↓ ↓ ÷10

$1 : 1.8$

This equivalent ratio is in the form $1 : n$. This is useful for calculations.

Simplest form

To write a ratio in its simplest form, find an equivalent ratio with the smallest possible whole number values.

Simplest form
$5 : 1$ $10 : 9$
$2 : 3 : 4$

Not simplest form
$10 : 2$ $1 : 0.9$
$1 : 1.5 : 2$

Worked example

Alexis and Nisha share a flat. They decide to split their phone bill in the ratio 3 : 5
Alexis pays £78. How much does Nisha pay? **(2 marks)**

$78 ÷ 3 = 26$

$26 × 5 = 130$

Nisha pays £130

Work out the value of one part of the ratio first. You could also use equivalent ratios to solve this problem. Work out how many 3s go into 78 to find the multiplier.

$3 : 5$

×26 ↓ ↓ ×26

$78 : 130$

Check it!
$£130 + £78 = £208$
Divide £208 in the ratio 3 : 5
$3 + 5 = 8$ parts in the ratio in total
$£208 ÷ 8 = £26$
$£26 × 3 = £78$ ✓

Worked example

A university course has 945 applicants for 126 places.

(a) Find the ratio of the number of applicants to the number of places. Give your ratio in the form $n : 1$ **(2 marks)**

$945 ÷ 126 = 7.5$

Applicants : Places $= 945 : 126$
$= 7.5 : 1$

(b) Of the 126 successful applicants, the ratio of males to females is 4 : 5
Work out the number of females on the course. **(2 marks)**

$4 + 5 = 9$
$126 ÷ 9 = 14$
$14 × 5 = 70$

There were 70 females on the course.

Work out the total number of parts in the ratio, then divide 126 by this to work out how many people each part represents. Then multiply by 5 to work out the number of females.

Now try this

Try to work this out without using your calculator.

1 Andre, Becky and Chrissie share some money in the ratios 3 : 6 : 7
In total Andre and Becky receive £207.
Work out the amount of money Chrissie receives. **(2 marks)**

2 The ratio of Amir's age to Petra's age is 3 : 7
Amir is 9 years old.
(a) Work out Petra's age. **(2 marks)**

The ratio of Karl's age to Yasmina's age is 5 : 2
The sum of their ages is 84.
(b) Work out Yasmina's age. **(2 marks)**

Proportion

Two quantities are in **direct proportion** when both quantities increase at the same rate.

Number of theatre tickets bought   Total cost

   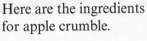

3   £135

×3   ×3

9   £405

Two quantities are in **inverse proportion** when one quantity increases at the same rate as the other quantity decreases.

Average speed   Time taken

40 km/h   2 hours

×2   ÷2

80 km/h   1 hour

Divide or multiply?

You can use common sense to work out whether to divide or multiply in proportion questions.

6 people can build a wall in 4 days.

6 × 4 = 24 so 1 person could build the wall in 24 days.

Multiply because it would take 1 person **longer** to build the wall.

24 ÷ 8 = 3 so 8 people could build the wall in 3 days.

Divide because it would take 8 people **less time** to build the wall.

Worked example

Here are the ingredients for apple crumble.

Apple crumble	
Serves 6 people	
900 g apples	90 g butter
180 g sugar	150 g flour

(a) Henry wants to make apple crumble for 11 people.
Work out the amount of sugar he needs. **(2 marks)**

Amount needed for 1 person = $\frac{180}{6}$ = 30 g

Amount needed for 11 people = 11 × 30 = 330 g

(b) Carla makes an apple crumble using 2250 g of apples. Work out how many people her apple crumble will serve. **(2 marks)**

2250 ÷ 900 = 2.5
6 × 2.5 = 15
The apple crumble will serve 15 people.

Problem solved!

There are two ways to tackle this question:
- Work out the cost per unit of volume for each option.
- Work out the amount of lemonade you get for a fixed amount of money for each option.

It doesn't matter which method you use, but you need to show how you arrived at your conclusion. The easiest way to do this is to write down **what you are working out** at each stage.

> You will need to use problem-solving skills throughout your exam – **be prepared!**

Worked example

Alice wants to buy some lemonade for a party. She compares the prices of a two-litre bottle and a multi-pack of eight cans.
- 2-litre bottle   £1.29
- Eight 330 ml cans   £1.99

Which option offers the best value? **(3 marks)**

Cost in pence per ml:
Two-litre bottle: 129 ÷ 2000 = 0.0645
Cans: 199 ÷ (8 × 330) = 0.075 37...
Lemonade costs less per ml in a two-litre bottle so that is the best value.

Now try this

Worked solution video

1.25 kg of cheese costs €16.55 in France.
$1\frac{1}{2}$ lb of the same cheese costs £8.97 in England.
In which country is it cheaper to buy the cheese?
Show all of your working. **(5 marks)**
1 kg = 2.2 lb   £1 = €1.15

> You need to write quantities in the same units before comparing them. Choose lb or kg and choose £ or €. Remember to write a conclusion.

Percentage change

There are two methods that can be used to increase or decrease an amount by a percentage.

Method 1

Work out 26% of £280:

$$\frac{26}{100} \times £280 = £72.80$$

£280

Subtract the decrease:

£280 − £72.80 = £207.20

26% OFF

Method 2

Use a multiplier:

100% + 30% = 130%

$$\frac{130}{100} = 1.3$$

So the multiplier for a 30% increase is 1.3:

400 g × 1.3 = 520 g

400 g PLUS **30% EXTRA**

Worked example

For the 2014/2015 season a football club charged £550 for a top-price season ticket.

For the 2015/2016 season the club increased its season ticket prices by 6.2%.

It offered supporters a 10% discount on the new price if they bought their season tickets before the end of June.

Calculate the price of a season ticket bought before the end of June. **(3 marks)**

550 × 1.062 = 584.1
584.1 × 0.9 = 525.69
The discounted price is £525.69

Problem solved!

Read the question carefully – there are two steps in the working:

6.2% increase
The multiplier is $\frac{100 + 6.2}{100} = 1.062$

10% decrease
The multiplier is $\frac{100 - 10}{100} = 0.9$

Make sure you give **units** with your final answer.

> You will need to use problem-solving skills throughout your exam – **be prepared!**

Calculating a percentage increase or decrease

Work out the amount of the increase or decrease

↓

Write this as a percentage of the original amount

Was £60 Now £39

60 − 39 = 21

$$\frac{21}{60} \times 100 = 35\%$$

This is a 35% decrease.

> For a reminder about writing one quantity as a percentage of another, have a look at page 59.

A question might ask you to calculate a percentage **profit** or **loss** rather than an increase or decrease.

Now try this

1 A TV originally cost £520.

In a sale it was priced at £340.

What was the percentage reduction in the price?

Give your answer to 1 decimal place. **(3 marks)**

> **Reduction** means decrease. Work out the decrease as a percentage of the original price.

2 Johan publishes a monthly poetry magazine. In March he printed 1400 copies of his magazine. In April he increased his print run by 15%.

It costs Johan £800 plus 75p per copy to print his magazine. He sells each issue for £1.99.

Assuming Johan sells every copy that he prints, calculate his percentage profit in April. **(4 marks)**

Reverse percentages

In some questions you are given an amount **after** a percentage change, and you have to find the **original amount**. To answer questions like this you need to be really confident with **percentage change**. Revise it on page 62.

Using a multiplier

You can use a multiplier to calculate a percentage increase or decrease. If you are given the **final amount** and you need to find the **original amount**, you can **divide by the multiplier**. Here are two examples.

 A sweater is **reduced** in price by 20% in a sale.

Original price £50 × 0.8 Sale price £40 ÷ 0.8

The average temperature **increases** by 5%.

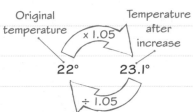

Original temperature 22° × 1.05 Temperature after increase 23.1° ÷ 1.05

Worked example

In a sale, normal prices are reduced by 15%
The sale price of a pair of trainers is £75.65
Work out the normal price of the trainers. **(3 marks)**

100% − 15% = 85%

$\frac{85}{100} = 0.85$

75.65 ÷ 0.85 = 89

The original price was £89

Read percentages questions carefully so you know what information you have been given. This question tells you the price **after** the percentage decrease, so you need to use **reverse percentages**. To find the multiplier for a 15% decrease:
1. Subtract 15% from 100%
2. Divide by 100 to convert to a multiplier.

You need to **divide** by the multiplier to find the original price.

Check it!
Reduce £89 by 15%: £89 × 0.15 = £13.35
£89 − £13.35 = £75.65 ✓

Problem solved!

Plan your answer before you start. You will need to do **two separate calculations** to find both original heights. Then you need to **compare** the original heights and **write a conclusion**.
To find a multiplier for a percentage increase:
1. add the percentage to 100%
2. divide by 100 to convert to a multiplier.
You need to **divide** by the multiplier to find the height in 2013.

You will need to use problem-solving skills throughout your exam – **be prepared!**

Worked example

Amy and Paul measure their heights each year. This table shows their heights in 2014.

	Height in 2014 (cm)	Percentage increase since 2013
Paul	154	10%
Amy	150.8	4%

Who was taller in 2013. Give reasons for your answer. **(4 marks)**

Paul's original height = 154 ÷ 1.1 = 140 cm
Amy's original height = 150.8 ÷ 1.04 = 145 cm
In 2013 Amy was taller than Paul by 5 cm.

Now try this

1 Hannah buys a pair of shoes in a sale where all the items are marked '40% off'.
She pays £27 for the shoes.
What price were the shoes originally? **(3 marks)**

2 Jared bought a house in 2010.
By 2012 his house had increased in value by 8%. The new value of Jared's house is £237 600. How much did Jared pay for his house? **(3 marks)**

Worked solution video

Growth and decay

You can use **repeated percentage change** to model problems involving growth and decay.

Compound interest.

If you leave your money in a savings account it will earn compound interest.

Saanvi invests £40 000 at a compound interest rate of 3% per annum.

'Per annum' means 'per year'

This table uses a multiplier to work out the balance of Saanvi's account at the end of each year.

End of year	Balance (£)
1	40 000 × 1.03 = 41 200
2	41 200 × 1.03 = 42 436
3	42 436 × 1.03 = 43 709.08

The multiplier for a 3% increase is × 1.03

You can use indices to work out the final balance after 3 years more easily.

Balance after 3 years
= £40 000 × 1.03 × 1.03 × 1.03
= £40 000 × 1.03³
= £43 709.08

Golden rule

You can use this rule to calculate a repeated percentage change:

Final amount = (starting amount) × (multiplier)^n

n is the number of times the change is made.

LEARN IT!

Worked example

At the start of an experiment a Petri dish contains 5000 cells. The number of cells in the Petri dish increases by 20% each day.

Calculate the number of cells in the Petri dish at the end of 4 days. **(2 marks)**

5000 × 1.2⁴ = 10 368

You can enter this in one go on your calculator by using the $x^\square$ button.

Worked example

A scientist is creating a model for the number of carp in a lake. Her model predicts that $C_{n+1} = 0.8C_n$ where C_n is the number of carp after n years.

(a) Does this model predict that the number of carp is increasing or decreasing each year? **(1 mark)**

Decreasing by 20% each year.

(b) $C_0 = 600$. Calculate (i) C_1 (ii) C_3 **(2 marks)**
(i) $C_1 = 0.8 × C_0 = 0.8 × 600 = 480$
(ii) $C_2 = 0.8 × C_1 = 0.8 × 480 = 384$
  $C_3 = 0.8 × C_2 = 0.8 × 384 = 307$ (nearest whole number)

This question uses iteration notation. There is more about this on page 45. The values C_0, C_1, C_2 and so on form a sequence that decreases with each term.

$C_n = 0.8^n × C_0$ ←Starting amount

Final amount   Multiple

(c) This model predicts that after k years the number of carp in the lake will have dropped below 200. Calculate the value of k. **(2 marks)**

After 4 years: $C_4 = 0.8^4 × 600 = 246$ (nearest whole number)
After 5 years: $C_4 = 0.8^5 × 600 = 197$ (nearest whole number)
$k = 5$

Now try this

1 A census predicts that the population of a city will decrease by 2% each year. At the end of 2014 the population of the city is 260 000.
Estimate the population of the city at the end of 2020 Give your answer to the nearest 100 people. **(2 marks)**

Worked solution video

2 Amir invests £5000 in a savings account. He is paid 3% per annum compound interest.
(a) How much will Amir have in his savings account after 2 years? **(2 marks)**
(b) Amir needs £5600 to buy a car. Calculate the number of years Amir would need to leave his money in the account to save up this amount. **(2 marks)**

Speed

This is the formula triangle for speed.

Average speed

Distance

Time

$$\text{Average speed} = \frac{\text{total distance travelled}}{\text{total time taken}}$$

$$\text{Time} = \frac{\text{distance}}{\text{average speed}}$$ **LEARN IT!**

$$\text{Distance} = \text{average speed} \times \text{time}$$

Using a formula triangle

Cover up the quantity you want to find with your finger.

The position of the other two quantities tells you the formula.

$$T = \frac{D}{S} \qquad S = \frac{D}{T} \qquad D = S \times T$$

Units

The most common units of speed are:

- metres per second: m/s
- kilometres per hour: km/h
- miles per hour: mph.

To convert between measures of speed you need to convert one unit first then the other. Write the new units at each step of your working. To convert 72 km/h into m/s:

$$72\,\text{km/h} \rightarrow 72 \times 1000 = 72\,000\,\text{m/h}$$

$$72\,000\,\text{m/h} \rightarrow 72\,000 \div 3600 = 20\,\text{m/s}$$

1 hour = 60 × 60 = 3600 seconds

Minutes and hours

For questions on speed, you need to be able to convert between minutes and hours.

Remember there are 60 minutes in 1 hour.

To convert from minutes to hours you divide by 60.

24 minutes = 0.4 hours $\frac{24}{60} = \frac{2}{5} = 0.4$

To convert from hours to minutes you multiply by 60.

0.2 hours = 12 minutes   3.2 × 60 = 192

3.2 hours = 3 hours 12 minutes

Worked example

The speed of light in a vacuum is approximately 1.08×10^9 km/h.

Light from the Sun takes approximately 8 minutes and 15 seconds to travel to Earth.

Estimate the distance from the Earth to the Sun.   **(3 marks)**

8 mins 15 secs = 8.25 mins = $\frac{8.25}{60}$

= 0.1375 hours

$D = S \times T$

= $1.08 \times 10^9 \times 0.1375$

= 1.485×10^8 km

Be careful with the units. You need to convert 12 minutes and 15 minutes into hours before doing your calculations.

Speed checklist

Draw formula triangle.   ✓

Make sure units match.   ✓

Give units with answer.   ✓

If you're answering questions involving speed, distance and time you must always make sure that the units match. Speed is given in km/h here, so convert the time into hours before calculating.

Now try this

Rosa lives in Durham and works in Newcastle. She takes the train to work every day.

Last Tuesday her train journey to work took 12 minutes, at an average speed of 108 km/h.

Her journey home from work took 15 minutes.

Calculate Rosa's average speed on her journey home.   **(3 marks)**

Density

The density of a material is its mass per unit volume.

This is the formula triangle for density.

$$\text{Density} = \frac{\text{mass}}{\text{volume}}$$

$$\text{Volume} = \frac{\text{mass}}{\text{density}}$$

$$\text{Mass} = \text{density} \times \text{volume}$$

Worked example

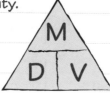

The diagram shows a solid hexagonal prism.
The area of the cross-section of the prism is $15\,\text{cm}^2$.
The length of the prism is 8 cm.
The prism is made from wood with a density of 0.8 grams per cm^3.
Work out the mass of the prism.   **(4 marks)**

Volume of prism

    $= \text{area of cross-section} \times \text{length}$

    $= 15 \times 8$

    $= 120\,\text{cm}^3$

$M = D \times V$

    $= 0.8 \times 120$

    $= 96$

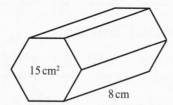

The mass of the prism is 96 g.

Revise volumes of prisms on page 82.

Units

The most common units of density are:

- grams per cubic centimetre: g/cm^3
- kilograms per cubic metre: kg/m^3

$\text{Mass} = \text{density} \times \text{volume}$

You are given the density so you need to work out the volume of the prism.

You need to learn the formula for the volume of a prism.

The density is in grams per cm^3 and the volume is in cm^3 so the mass will be in grams.

Worked example

An iron bar has a volume of $1.2\,\text{m}^3$ and a mass of 9444 kg. Calculate the density of iron.   **(2 marks)**

$$D = \frac{M}{V} = \frac{9444}{1.2} = 7870\,\text{kg}/\text{m}^3$$

Volume is in m^3 and mass is in kg so density will be in kg/m^3.

Worked solution video

Now try this

The density of copper is $8.92\,\text{g}/\text{cm}^3$.
The density of silver is $10.49\,\text{g}/\text{cm}^3$.
$20\,\text{cm}^3$ of copper and $5\,\text{cm}^3$ of silver are mixed together to make a new kind of metal.
Work out the density of the new metal.
          (4 marks)

Plan your answer:
1. Work out the mass of each metal.
2. Add the masses together.
3. You know the total mass and the total volume of the new metal, so calculate the density.

Other compound measures

Compound measures are made up of two or more other measurements. **Speed** is a compound measure because it is calculated using distance **and** time. **Density** is a compound measure because it is calculated using mass **and** volume. You need to be able to work with other compound measures as well.

Pressure

Pressure is a measure of the force applied over a given area. The most common units of pressure are Newtons per square centimetre (N/cm^2) and Newtons per square metre (N/m^2). You can use the formula triangle on the right to calculate with pressure.

$$Pressure = \frac{Force}{Area}$$

$$Area = \frac{Force}{Pressure}$$

$$Force = Pressure \times Area$$

LEARN IT!

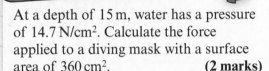

Worked example

At a depth of 15 m, water has a pressure of 14.7 N/cm^2. Calculate the force applied to a diving mask with a surface area of 360 cm^2.　　**(2 marks)**

Force = Pressure × Area
　　= 14.7 × 360
　　= 5292 N

Rates

If the 'bottom' unit is in a compound measure **time**, then it is a **rate**. Here are some examples.

$$Speed = \frac{Distance}{Time}$$ 　　$$Rate\ of\ flow = \frac{Volume}{Time}$$

$$Rate\ of\ climb = \frac{Height}{Time}$$ 　　$$Rate\ of\ pay = \frac{Salary}{Time}$$

Problem solved!

You need to use lots of different maths skills to solve this problem. You need to know how to calculate the volume of a cuboid and how to convert from cm^3 to litres (for a reminder of both look at page 81). It's a good idea to plan your strategy before you start:

1. Work out the volume of the fish tank.

2. Convert from cm^3 into litres.

3. Divide by 2 to get half the capacity.

4. Divide by the rate of flow to get the time taken in minutes.

> You will need to use problem-solving skills throughout your exam – **be prepared!**

Worked example

This fish tank can be modelled as a cuboid.

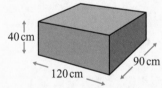

40 cm
90 cm
120 cm

Aaron uses a hose to fill the fish tank with water at a rate of 12 litres per minute. How long will it take for the fish tank to be half full?　　**(5 marks)**

Volume = 120 × 90 × 40
　　= 432 000 cm^3
　　= 432 litres
432 ÷ 2 = 216
216 ÷ 12 = 18
It will take 18 minutes for the tank to be half full.

Now try this

1 The average fuel consumption of a car is measured in kilometres per litre (km/l). A car travels 249 km and uses 15 litres of petrol. What is its average fuel consumption?　　**(2 marks)**

Look at the units to work out what calculation to do.

2 A large tank holds 1680 litres of water. The tank can be filled from a hot tap or from a cold tap. The cold tap on its own takes 4 minutes to fill the tank. The hot tap on its own takes 6 minutes to fill the tank.

Preti turns both taps on at the same time. How long does the tank take to fill?　　**(3 marks)**

Proportion and graphs

You can use the symbol ∝ to show direct proportion. You can show quantities that are **directly proportional** or **inversely proportional** on a graph.

Direct proportion facts

If y is directly proportional to x:

☑ you can write $y \propto x$

☑ you can write an equation $y = kx$ where k is a number

☑ the graph of x against y is a **straight line** passing **through the origin**.

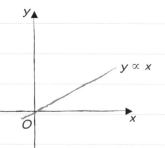

$y \propto x$

Inverse proportion facts

If y is inversely proportional to x:

☑ you can write $y \propto \dfrac{1}{x}$ *y is directly proportional to the **reciprocal** of x*

☑ you can write an equation $y = \dfrac{k}{x}$ where k is a number

☑ the graph of x against y looks like a **reciprocal graph**.

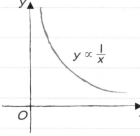

$y \propto \dfrac{1}{x}$

Worked example

p and q are inversely proportional. Circle the equation that could describe the relationship between p and q.

$p = 2q$ $p = q + 5$ $p = \dfrac{q}{10}$ $\boxed{p = \dfrac{2}{q}}$ **(1 mark)**

> An equation for **inverse** proportionality looks like $y = \dfrac{k}{x}$ where k is a number.

There are other ways to answer this question. The method used here is a bit like using **equivalent ratios**:

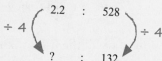

$\div 4 \begin{pmatrix} 2.2 &:& 528 \\ ? &:& 132 \end{pmatrix} \div 4$

Check it!
Your answer should make sense. If the quantities are in **direct proportion** then when one decreases the other should decrease as well. ✓

Worked example

Electrical appliances in your home follow this rule:

Power (watts) ∝ Current (amps)

An electric drill uses 2.2 A and has a power of 528 W.
Calculate the current used by a television with a power of 132 W.   **(2 marks)**

$528 \div 132 = 4$
$2.2 \div 4 = 0.55$
The television uses 0.55 A

Now try this

This graph can be used to convert between inches and centimetres.

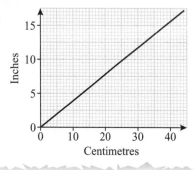

 (a) Use the graph to convert 10 cm to inches.   **(1 mark)**

(b) Use the graph to convert 12.5 inches to cm.   **(1 mark)**

(c) What evidence is there from the graph to show that inches are directly proportional to centimetres.   **(2 marks)**

Proportionality formulae

You can answer some questions involving proportion by constructing a **formula**. On this page you can revise finding formulae for the two basic proportionality relationships.

1 ## Direct proportion
These all mean the same thing:

- y is directly proportional to x
- y varies directly with x
- $y \propto x$
- $y = kx$ — k is called the **constant** of proportionality.

2 ## Inverse proportion
These all mean the same thing:

- y is inversely proportional to x
- y varies inversely with x
- $y \propto \dfrac{1}{x}$
- $y = \dfrac{k}{x}$

Worked example

Winnie drops a stone down a well. The speed of the stone, v m/s, is directly proportional to the time, t seconds, since she dropped it.
After 0.5 seconds the stone is travelling at 4.9 m/s.

(a) Find a formula for v in terms of t. **(3 marks)**

$v = kt$
$4.9 = k(0.5)$ $(\div 0.5)$
$k = 9.8$
$v = 9.8t$

(b) Calculate the speed of the stone after 1.2 seconds. **(1 mark)**

$v = 9.8(1.2) = 11.76$ m/s

You can find a proportionality formula if you know the type of proportionality and are given two corresponding values for the variables. Follow these steps:

1. Write down the formula using k for the constant of proportionality.
2. Substitute the values of v and t you are given.
3. Solve the equation to find the value of k.
4. Write down the formula putting in the value of k.
5. Once you have written your formula, you can use it to find the value of one variable if you know the value of the other.

Checking for proportionality

You can use a graph to check whether two quantities are directly proportional.

P	5	10	15
Q	1.2	1.5	1.8

The graph doesn't go through the origin so P and Q are not directly proportional.

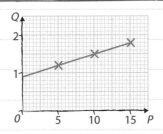

Now try this

1 x is directly proportional to y.
When $x = 36$, $y = 5$

(a) Find a formula for x in terms of y. **(3 marks)**

(b) Calculate the value of x when $y = 32$ **(1 mark)**

(c) Calculate the value of y when $x = 9$ **(1 mark)**

Worked solution video

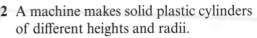

2 A machine makes solid plastic cylinders of different heights and radii.

The height h cm of a plastic cylinder is inversely proportional to its radius r cm.

A plastic cylinder of height 6 cm has a radius of 4 cm.

Work out the height of a plastic cylinder with a radius of 3 cm. **(4 marks)**

Harder relationships

You can use proportion to describe harder relationships between variables. Here are some relationships and graphs you should be familiar with.

Proportionality in words	Using $\propto$	Formula
y is directly proportional to the square of x	$y \propto x^2$	$y = kx^2$
y is directly proportional to the cube of x	$y \propto x^3$	$y = kx^3$
y is directly proportional to the square root of x	$y \propto \sqrt{x}$	$y = k\sqrt{x}$
y is inversely proportional to the square of x	$y \propto \dfrac{1}{x^2}$	$y = \dfrac{k}{x^2}$

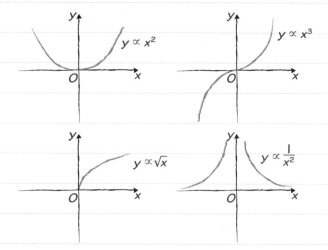

Worked example

q varies inversely with the square of t.

(a) What happens to q if t is doubled? **(1 mark)**

It is divided by 4.

(b) When $t = 4$, $q = 8.5$
Calculate the value of q when $t = 5$

(4 marks)

$q \propto \dfrac{1}{t^2}$

$q = \dfrac{k}{t^2}$

$8.5 = \dfrac{k}{4^2}$

$k = 8.5 \times 4^2 = 136$

$q = \dfrac{136}{t^2}$

When $t = 5$: $q = \dfrac{136}{5^2} = 5.44$

'q varies inversely with the square of t' means 'q is inversely proportional to t^2'.

$q = \dfrac{k}{t^2}$ so if t is doubled the new value of q is:

$\dfrac{k}{(2t)^2} = \dfrac{k}{4t^2} = \tfrac{1}{4}q$

Always...

1. Write down the statement of proportionality and then the formula.
2. Substitute the values you are given.
3. Solve the equation to find k.
4. Write down the formula using the value of k.
5. Use your formula to find any unknown values.

Now try this

1 A stone is dropped off a cliff. It takes t seconds to fall a distance d metres.

t is directly proportional to $\sqrt{d}$

When $t = 4.6$, $d = 25$

(a) Express t in terms of d. **(3 marks)**

(b) Find the length of time the stone takes to fall 42.25 m. **(2 marks)**

2 The speed v (km/s) of a satellite is inversely proportional to the square root of the radius of its orbit r (km).

The Hubble space telescope orbits at a radius of 6940 km at a speed of 7.6 km/s.

Work out the speed of a satellite orbiting at a radius of 42 000 km. Give your answer to 3 significant figures. **(4 marks)**

Worked solution video

Problem-solving practice 1

Throughout your Higher GCSE exam you will need to **problem-solve, reason, interpret** and **communicate** mathematically. If you come across a tricky or unfamiliar question in your exam you can try some of these strategies:

✓ Sketch a diagram to see what is going on.

✓ Try the problem with smaller or easier numbers.

✓ Plan your strategy before you start.

✓ Write down any formulae you might be able to use.

✓ Use *x* or *n* to represent an unknown value.

1 At a sixth form college, the ratio of male to female students is 7 : 6

There are 144 female students at the college. How many students are there in total? **(2 marks)**

Ratio page 60

There is more than one strategy here. One approach would be to work out what one part of the ratio represents.

TOP TIP

Check your answer by dividing it in the ratio 7 : 6

2 Suresh invested £2000 in a savings account for two years.

The account pays 2% per annum compound interest.

At the end of the first year he put an extra £1500 into the account.

How much money will Suresh have at the end of the two years? **(4 marks)**

Growth and decay page 64

Work out the interest Suresh earns on £2000, then add the extra £1500. This is the new starting amount for the second year. So in the second year, Suresh earns interest on this new amount.

TOP TIP

At each step of your working, make a note of what you have worked out. Your notes here could be:
• Interest earned in year 1
• Total starting amount for year 2
• Interest earned in year 2
• Total amount at end of year 2

3 Anton is driving to visit his sister. He travels 90 miles at an average speed of 40 mph.

He then increases his average speed to 60 mph for a further 90 miles. Calculate his average speed for the entire journey. **(3 marks)**

Speed page 65

Watch out! The answer is not 50 mph. Think about what information you need to calculate his average speed.

TOP TIP

The number of marks can give you a clue as to how much work is involved in a question. If the question is worth 3 marks you probably need to do more than one simple calculation.

Problem-solving practice 2

4 This item appeared in a newspaper.

> **Cow produces 3% more milk**
> A farmer found that when his cow listened to classical music the milk it produced increased by 3%.
> This increase of 3% represented 0.72 litres of milk.

Calculate the amount of milk produced by the cow when it listened to classical music.

(3 marks)

Proportion page 61
Reverse percentages page 63

When the cow listened to classical music, it produced 103% of the milk it produced originally. You know that 3% represents 0.72 litres. Use this information to work out what 103% represents.

TOP TIP

You can sometimes solve percentage problems by working out what 1% represents.

---

5 A paint manufacturer mixes white and green paint in two different ratios to make different shades.

Rosa has 2 litres of Summer Fields. How much extra white paint should she add to turn it into Apple White? **(4 marks)**

Ratio page 60

Start by working out how much white and green paint are contained in the 2 litres of Summer Fields that Rosa already has. Her new mixture will contain the same amount of **green** paint. You can use the ratio to work out what the **total** amount of white paint should be in the new mixture.

TOP TIP

Plan your strategy before you start – you will save time and your working will be much clearer.

---

6 The time taken for a clock pendulum to swing once is proportional to the square root of its length.

In the diagram below, pendulum B takes 0.8 seconds longer to swing than pendulum A.

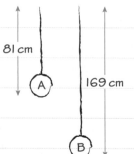

A clock-maker wants a pendulum that takes 1 second to swing. Find the length of this pendulum. **(6 marks)**

Harder relationships page 70

Be careful! You don't know how long either pendulum will take to swing – only the **difference** between the times. Use this information to write a pair of simultaneous equations and solve them to find k. Then use the proportionality equation to find the length for a time of 1 second.

TOP TIP

You can answer lots of questions involving proportionality using an equation. If time (T) is proportional to the square root of length (L) then you can write $T = k\sqrt{L}$

Angle properties

You need to remember all of these angle properties and their correct names.

Corresponding angles are equal.

Alternate angles are equal.

Allied angles (or **co-interior angles**) add up to 180°.

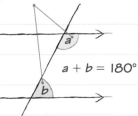

$a + b = 180°$

Parallel lines are marked with arrows.

Vertically opposite angles are equal.

These are useful angle facts for triangles and parallelograms:

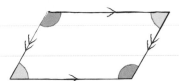

Interior angle

The exterior angle of a triangle is equal to the sum of the interior angles at the other two vertices.

Exterior angle

The opposite angles of a parallelogram are equal.

You need to know the proofs of the angle properties of triangles and quadrilaterals.

Golden rule

When answering angle problems, you need to give a reason for each step of your working.

Angle sums

You need to remember these two angle facts:

1 The angles in a triangle add up to 180°.

2 The angles in a quadrilateral add up to 360°.

Worked example

Use the fact that the angles in a triangle add up to 180° to write an equation, then solve your equation to find x. For a reminder about solving linear equations have a look at page 19.

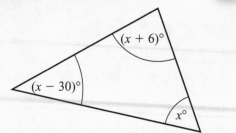

$(x + 6)°$

$(x - 30)°$

$x°$

Work out the value of x.   **(3 marks)**

Angles in a triangle add up to 180° so

$(x - 30) + (x + 6) + x = 180$

$3x - 24 = 180 \quad (+ 24)$

$3x = 204 \quad (÷ 3)$

$x = 68$

Now try this

AB is parallel to CD.

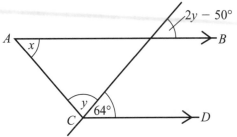

$2y - 50°$

$A \quad x \quad \longrightarrow B$

$y \quad 64°$

$C \quad \longrightarrow D$

Work out the value of x.   **(5 marks)**

Solving angle problems

You might need to use angle properties to solve problems in your exam.
Remember to give reasons for every step of your working.

Reasons

Use these reasons in angle problems:
- Angles on a straight line add up to 180°.
- Angles around a point add up to 360°.
- Opposite angles are equal.
- Corresponding angles are equal.
- Co-interior angles add up to 180°.
- Alternate angles are equal.
- Angles in a triangle add up to 180°.
- Angles in a quadrilateral add up to 360°.
- Base angles of an isosceles triangle are equal.

Use the properties on the diagram:
AB is parallel to CD
AC is parallel to BD
BE is equal to DE

Worked example

Work out the size of the angle marked x.
Give reasons for each step of your working.

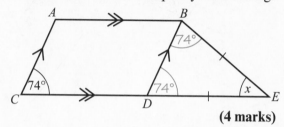

(4 marks)

$\angle BDE = 74°$ (corresponding angles are equal)

$\angle DBE = 74°$ (base angles in an isosceles triangle are equal)

$x + 74° + 74° = 180°$ (angles in a triangle add up to 180°)

$x = 180° - 148°$

$x = 32°$

Worked example

ABCD is a quadrilateral.
Angle DAB is a right angle.
Angles ABC and BCD are in the ratio 1 : 2
Angle CDA is 70° more than angle ABC.
Work out the size of angles ABC, BCD and CDA.

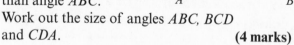

(4 marks)

$A = 90°$

$C = 2x$

$D = x + 70°$

$A + B + C + D = 360°$ (angles in a quadrilateral add up to 360°)

So $90° + x + 2x + (x + 70°) = 360°$

$4x = 200°$

$x = 50°$

So $B = 50°$, $C = 2 \times 50° = 100°$ and
$D = 50° + 70° = 120°$

Write angle B as x then express angles C and D in terms of x.
Angle problems could involve ratio or proportion, or you might need to write your own equation.

Now try this

In the diagram ABC and BDC are isosceles triangles.
Express the size of angle ABD in terms of x, giving your answer as simply as possible.
Give a reason for each step of your working.

(4 marks)

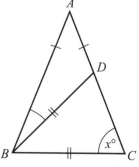

Angles in polygons

Polygon questions are all about interior and exterior angles.

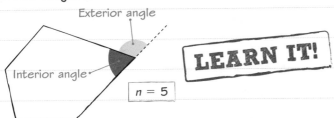

Exterior angle

Interior angle

$n = 5$

LEARN IT!

Use these formulae for a polygon with n sides.

Sum of interior angles $= 180° \times (n - 2)$

Sum of exterior angles $= 360°$

This diagram shows part of a **regular** polygon with 30 sides.

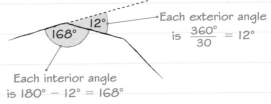

168° 12°

Each exterior angle is $\frac{360°}{30} = 12°$

Each interior angle is $180° - 12° = 168°$

Don't try to draw a 30-sided polygon!

If there's no diagram given in a polygon question, you probably don't need to draw one.

Regular polygons

In a regular polygon all the sides are equal and all the angles are equal.

If a regular polygon has n sides then each exterior angle is $\frac{360°}{n}$

LEARN IT!

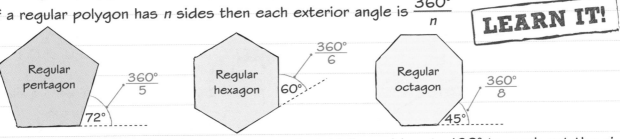

Regular pentagon $\frac{360°}{5}$ 72°

Regular hexagon 60° $\frac{360°}{6}$

Regular octagon $\frac{360°}{8}$ 45°

You can use the fact that the angles on a straight line add up to $180°$ to work out the size of one of the interior angles.

Worked example

The diagram shows part of a regular polygon. The interior angle and the exterior angle at a vertex are marked.

The size of the interior angle is 7 times the size of the exterior angle.

Work out the number of sides of the polygon. **(3 marks)**

$180° \div 8 = 22.5°$

$\frac{360°}{22.5°} = 16$

The polygon has 16 sides.

Problem solved!

It's usually easier to work with **exterior** angles in polygon questions. You can rearrange the formula for the size of an exterior angle to get:

$$n = \frac{360°}{\text{exterior angle}}$$

You can use ratios to answer this question. The ratio of the interior angle to the exterior angle is $7 : 1$. These angles add up to $180°$ so divide $180°$ in the ratio $7 : 1$ to find the exterior angle.

You will need to use problem-solving skills throughout your exam – **be prepared!**

Now try this

The diagram shows part of a regular polygon with n sides.

(a) Work out the value of n. **(2 marks)**

(b) What is the sum of the interior angles of the polygon? **(2 marks)**

12°

Pythagoras' theorem

Pythagoras' theorem is a really useful rule. You can use it to find the length of a missing side in a right-angled triangle.

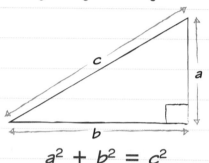

$$a^2 + b^2 = c^2$$

Pythagoras checklist

short² + short² = long² ✓

Right-angled triangle. ✓

Lengths of two sides known. ✓

Length of third side missing. ✓

Learn this. ✓

Worked example

This right-angled triangle has sides x, 17 cm and 8 cm.

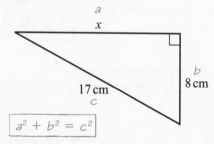

$$a^2 + b^2 = c^2$$

Show that $x = 15$ cm   **(2 marks)**

$x^2 + 8^2 = 17^2$

$\quad x^2 = 17^2 - 8^2$

$\quad\quad = 225$

$\quad x = \sqrt{225} = 15$ cm

The question says 'Show that' so you have to show **all** your working. Be careful when the missing length is one of the **shorter** sides.

1. Label the longest side of the triangle c.
2. Label the other two sides a and b.
3. Write out the formula for Pythagoras' theorem.
4. Substitute the values for a, b and c into the formula.
5. Rearrange the formula and solve. Make sure you show **every step** in your working.
6. Write units in your answer.

Pythagoras questions come in lots of different forms. Just look for the right-angled triangle.

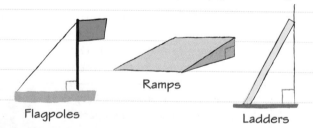

Flagpoles   Ramps   Ladders

Calculator skills

Use these buttons to find squares and square roots with your calculator.

x^2 $\sqrt{\square}$

You might need to use the S⇔D key to get your answer as a decimal number.

Now try this

(a) Work out the value of y.   **(2 marks)**

(b) Use your value of y to work out the value of z.   **(2 marks)**

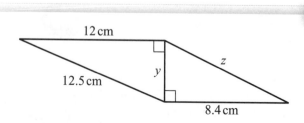

Trigonometry 1

You can use the trigonometric ratios to find the size of an angle in a right-angled triangle. You need to know the lengths of two sides of the triangle.

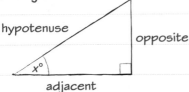

The sides of the triangle are labelled relative to the **angle** you need to find.

Trigonometric ratios **LEARN IT!**

$\sin x° = \dfrac{opp}{hyp}$ (remember this as S^O_H)

$\cos x° = \dfrac{adj}{hyp}$ (remember this as C^A_H)

$\tan x° = \dfrac{opp}{adj}$ (remember this as T^O_A)

You can use $S^O_H C^A_H T^O_A$ to remember these rules for trig ratios.

These rules only work for **right-angled** triangles.

Worked example

Calculate the size of angle x.   **(3 marks)**

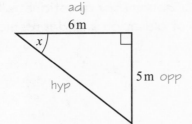

$\tan x° = \dfrac{opp}{adj} = \dfrac{5}{6}$

$x° = 39.805\,571\,09° = 39.8°$ (to 3 s.f.)

Label the **hyp**otenuse first — it's the longest side.

Then label the side **adj**acent to the angle you want to work out.

Finally, label the side **opp**osite the angle you want to work out.

Remember $S^O_H C^A_H T^O_A$. You know **opp** and **adj** here so use T^O_A.

Do **not** 'divide by tan' to get x on its own. You need to use the $\tan^{-1}$ function on your calculator.

$\tan^{-1}\left(\dfrac{5}{6}\right)$
        39.80557109

Write down all the figures on your calculator display then round your answer.

Using your calculator

To find a missing angle using trigonometry you have to use one of these functions.

$$\sin^{-1} \quad \cos^{-1} \quad \tan^{-1}$$

These are called **inverse trigonometric** functions. They are the inverse operations of sin, cos and tan.

Make sure that your calculator is in degree mode. Look for the ▨ symbol at the top of the display.

Now try this

Work out the size of angle x in each of these triangles. Give your answers correct to 1 decimal place.

(a)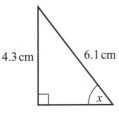

(3 marks)

(b)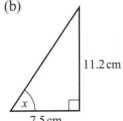

(3 marks)

(c)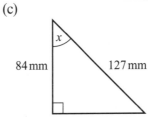

(3 marks)

Trigonometry 2

You can use the trigonometric ratios to find the length of a missing side in a right-angled triangle. You need to know the length of another side and the size of one of the acute angles.

Worked example

Calculate the length of side a. **(3 marks)**

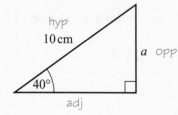

$$\sin x° = \frac{opp}{hyp}$$

$$\sin 40° = \frac{a}{10}$$

$$a = 10 × \sin 40°$$

$$= 6.427\,87...$$

$$= 6.43\,cm \text{ (to 3 s.f.)}$$

Label the sides of the triangle relative to the 40° angle. Write $S^O_H C^A_H T^O_A$ and tick the pieces of information you have. You need to use S^O_H here.

Write the values you know in the rule and replace **opp** with a. You can solve this equation to find the value of a.

Write down at least four figures of the calculator display before giving your final answer correct to 3 significant figures.

Check it!

Side a must be shorter than the hypotenuse. 6.43 cm looks about right. ✓

Angles of elevation and depression

Some trigonometry questions will involve angles of elevation and depression.

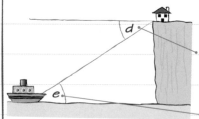

The angle of depression of the ship from the house.

The angle of elevation of the house from the ship.

Angles of elevation and depression are always measured from the horizontal.

In this diagram, $d = e$ because they are alternate angles.

Now try this

In part (c) a is the hypotenuse. It will be on the bottom of the fraction when you substitute, so be careful with your calculation.

Work out the length of side a in each of these triangles.
Give your answers correct to 1 decimal place.

(a)
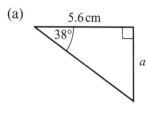

(b)
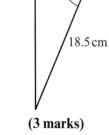

(c)
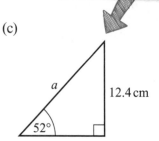

(3 marks)   **(3 marks)**   **(3 marks)**

Solving trigonometry problems

Non-calculator methods

You might need to answer questions involving **sin**, **cos** and **tan** without a calculator. Make sure you are confident with pages 77 and 78 before having a look at this page. Here are three sets of trigonometry values you need to **learn**.

① This triangle shows you the values of sin, cos and tan for 30° and 60°.

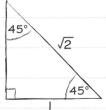

$$\sin 30° = \frac{1}{2} \qquad \sin 60° = \frac{\sqrt{3}}{2}$$

$$\cos 30° = \frac{\sqrt{3}}{2} \qquad \cos 60° = \frac{1}{2}$$

$$\tan 30° = \frac{1}{\sqrt{3}} \qquad \tan 60° = \sqrt{3}$$

② This triangle shows you the values of sin, cos and tan for 45°.

$$\sin 45° = \frac{1}{\sqrt{2}}$$

$$\cos 45° = \frac{1}{\sqrt{2}} \qquad \tan 45° = 1$$

LEARN IT!

③ You also need to know the values for 0° and 90°

$$\sin 0° = 0 \qquad \sin 90° = 1$$

$$\cos 0° = 1 \qquad \cos 90° = 0$$

$$\tan 0° = 0$$

tan 90° is **undefined**. If you enter it into a calculator you get an error.

You might get a question like this in your **non-calculator** paper, so make sure you learn the values of cos, sin and tan given above. If a triangle is **described** like this make sure you **sketch** it before doing any working.

Triangle *ABC* has a right angle at B. Angle *BAC* = 60°. *AC* = 14 cm. Calculate the length of *AB*. **(3 marks)**

$$\cos x° = \frac{adj}{hyp}$$

$$\cos 60° = \frac{AB}{14}$$

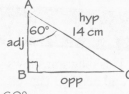

$$AB = 14 × \cos 60°$$

$$= 14 × \frac{1}{2} = 7 \text{ cm}$$

Amy arranges identical triangles like this around a point to make a pattern.

Show that she can fit exactly 12 triangles around a point. **(4 marks)**

$$\sin x° = \frac{opp}{hyp} = \frac{3}{6} = \frac{1}{2}$$

So $x° = 30°$
$3 × 12 = 36$
so $30 × 12 = 360$
There are 360 degrees around a point. 30 × 12 = 360, so she can fit exactly 12 triangles around a point.

Everything in red is part of the answer.

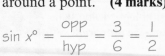

You need to be able to answer this question **without** a calculator.

This is a non-calculator question, so you will need to use your knowledge of exact trigonometric values. You need to know what the **inside** angle is, so label it *x*. The triangle is right-angled and you know two sides so you can use trigonometry. Make sure you **write a conclusion** explaining how your working shows the answer.

You will need to use problem-solving skills throughout your exam – **be prepared!**

A vertical flagpole *AB* is supported with a wire *AC* at an angle of 60° to the ground. The base of the wire is 2.4 m from the base of the flagpole. Calculate the length of the wire *AC*. **(3 marks)**

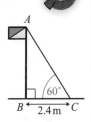

Perimeter and area

Triangle

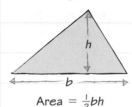

Area = $\frac{1}{2}bh$

Parallelogram

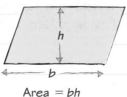

Area = bh

Trapezium

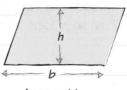

Area = $\frac{1}{2}(a + b)h$

LEARN IT!

You can calculate areas and perimeters of more complex shapes by splitting them into parts.

You might need to draw some extra lines on your diagram and add or subtract areas.

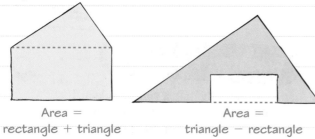

Area = rectangle + triangle

Area = triangle − rectangle

Area basics

Lengths are all in the same units.	✓
Give units with the answer.	✓
Lengths in cm means area units are cm².	✓
Lengths in m means area units are m².	✓

Worked example

This diagram shows a trapezium with base 6 cm and height x cm.

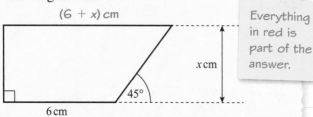

$(6 + x)$ cm

x cm

45°

6 cm

Everything in red is part of the answer.

The area of the trapezium is A cm².

(a) Write a formula for A in terms of x. Give your answer in its simplest form. **(3 marks)**

Area = $\frac{1}{2}(a + b)h$

$A = \frac{1}{2}(6 + 6 + x)x = 6x + \frac{1}{2}x^2$

(b) The height of the trapezium is 5 cm. Find the area of the trapezium. **(2 marks)**

$A = 6 \times 5 + \frac{1}{2} \times 5^2 = 42.5$ cm²

Problem solved!

You will probably have to solve unfamiliar problems involving area and perimeter. Here are some top tips:

- If you need to use a formula write it out before you substitute any values.
- Write any lengths or angles you work out neatly on the diagram given.
- Draw your own sketch if that helps.
- Don't be afraid to use algebra – you can substitute variables into a formula as well as numbers.
- If it's tricky, try using some numbers instead of variables first to see what is going on.

In this question you have to spot that the 45° angle makes it easy to work out the length of the top of the trapezium.

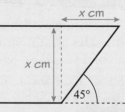

x cm

x cm

45°

You will need to use problem-solving skills throughout your exam – **be prepared!**

Now try this

The diagram shows a triangle.
In the diagram, all the measurements are in centimetres.
The perimeter of the triangle is 90 cm.
The area of the triangle is A cm².
Work out the value of A. **(4 marks)**

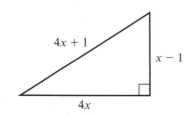

$4x + 1$

$x - 1$

$4x$

Worked solution video

Units of area and volume

Converting units of area or volume is trickier than converting units of length. You need to remember your area and volume conversions for your exam.

These two squares have the same area.

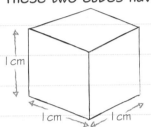

So $1\,cm^2 = 100\,mm^2$.

These two cubes have the same volume.

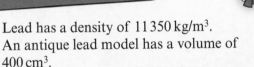

So $1\,cm^3 = 1000\,mm^3$.

Area conversions

$1\,cm^2 = 10^2\,mm^2 = 100\,mm^2$

$1\,m^2 = 100^2\,cm^2 = 10\,000\,cm^2$

$1\,km^2 = 1000^2\,m^2 = 1\,000\,000\,m^2$

Volume conversions

$1\,cm^3 = 10^3\,mm^3 = 1000\,mm^3$

$1\,m^3 = 100^3\,cm^3 = 1\,000\,000\,cm^3$

1 litre $= 1000\,cm^3$

$1\,ml = 1\,cm^3$

LEARN IT!

Worked example

Lead has a density of $11\,350\,kg/m^3$.
An antique lead model has a volume of $400\,cm^3$.
Calculate the mass of the model in kg.

(3 marks)

$400 \div 100^3 = 0.0004$

Volume $= 0.0004\,m^3$

Mass $=$ Density $\times$ Volume

$= 11\,350 \times 0.0004$

$= 4.54\,kg$

Unit conversion checklist

The multiplier for an area conversion is the length multiplier squared. ✓

The multiplier for a volume conversion is the length multiplier cubed. ✓

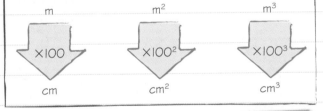

Problem solved!

You need to be really careful with the **units** when you are solving any question involving measures. The units of density are given in kg/m^3, so you need to convert $400\,cm^3$ into m^3 before you calculate. You are converting to a larger unit, so **divide** by 100^3.

For a reminder about density look at page 66.

> You will need to use problem-solving skills throughout your exam – **be prepared!**

Now try this

Worked solution video

1 Convert
   (a) $2.3\,m^2$ into cm^2 **(1 mark)**
   (b) $400\,mm^3$ into cm^3 **(1 mark)**

2 Convert $0.35\,m^3$ to mm^3, giving your answer in standard form.   **(2 marks)**

> For a reminder about writing answers in standard form have a look at page 8.

3 Jenny applies a force of $600\,N$ to the floor. The total area of her feet is $160\,cm^2$. What is the pressure, in N/m^2, between her and the floor if she stands on both of her feet?   **(3 marks)**

Prisms

Volume

A prism is a 3D solid with a **constant** cross-section. Use this formula to calculate the volume of a prism.

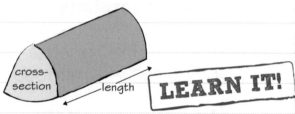

Volume = area of cross-section × length

Worked example

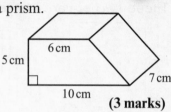

The diagram shows a prism. The cross-section is a trapezium.

Work out the volume of the prism.

(3 marks)

Area of cross-section (trapezium)
$$= \frac{1}{2} \times (6 + 10) \times 5 = 40\,cm^2$$
Volume of prism = $40 \times 7 = 280\,cm^3$

Surface area

To work out the surface area of a 3D shape, you need to add together the areas of all the faces.

It's a good idea to sketch each face with its dimensions.

Remember to include the faces that you can't see.

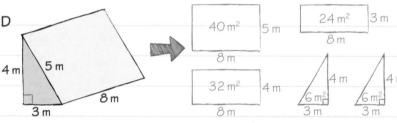

Surface area = $40 + 32 + 24 + 6 + 6 = 108\,m^2$

Worked example

The diagram shows a triangular prism and a cube. They both have the **same** volume.

Work out the length of x. **(4 marks)**

Volume of cube= $9^3 = 729\,cm^3$

Volume of prism = Area of cross-section × length
$$= \frac{1}{2} \times 5 \times 12 \times x$$
$$= 30x$$
$$30x = 9 \times 9 \times 9 = 729$$
$$x = 24.3\,cm$$

Problem solved!

Calculate the volume of the cube, and write an expression for the volume of the prism. Set these equal to each other and solve the equation to find x.

> You will need to use problem-solving skills throughout your exam – **be prepared!**

Use this formula to work out the areas of the triangular faces:

Area of a triangle = $\frac{1}{2}$ × base × vertical height

Now try this

The cross-section of this prism is a right-angled triangle.

(a) Work out the volume of the prism. **(3 marks)**

(b) Work out the total surface area of the prism. **(3 marks)**

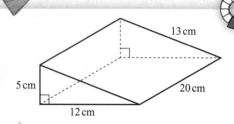

Circles and cylinders

You need to learn these formulae for circles and cylinders. They are not given in the exam.

Circle

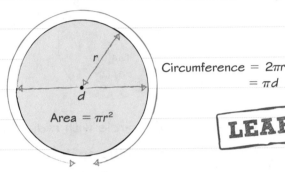

Circumference $= 2\pi r$
$= \pi d$

Area $= \pi r^2$

Cylinder

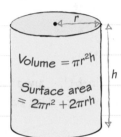

Volume $= \pi r^2 h$

Surface area $= 2\pi r^2 + 2\pi rh$

LEARN IT!

Worked example

The diagram shows a game counter in the shape of a semicircle.

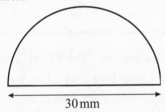

30 mm

Work out the area of the counter. Give your answer correct to 2 significant figures. **(3 marks)**

Radius $= 30 \div 2 = 15$ mm
Area of circle $= \pi r^2$
$= \pi \times 15^2$
$= 706.8583...$ mm^2
Area of counter $= 706.8583... \div 2$
$= 350$ mm^2 (2 s.f.)

Calculator skills

Make sure you know how to enter π on your calculator. On some calculators you have to press these keys:

SHIFT ➡ ×10^x π e

Problem solved!

The formula for the area of a circle uses the **radius**. If the length shown on the diagram is the **diameter**, you need to divide it by 2 before you substitute into the formula. Don't round any values until the end of your working.

You will need to use problem-solving skills throughout your exam – **be prepared!**

In terms of π

Unless a question asks you for a specific degree of accuracy, you can give your answers as a whole number or fraction multiplied by π. An answer given in terms of π is an **exact answer** rather than a **rounded answer**.

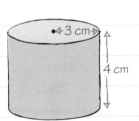

3 cm

4 cm

Volume of cylinder $= \pi r^2 h$
$= \pi \times 3^2 \times 4$

Exact answer
Volume $= 36\pi$ cm^3

Rounded answer
Volume $= 113$ cm^3 (to 3 s.f.)

Now try this

This shape is made from a rectangle and a semicircle.
Work out the area of the shape.
Give your answer correct to 3 significant figures.   **(4 marks)**

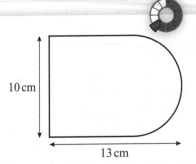

10 cm

13 cm

Sectors of circles

Each pair of radii divides a circle into two sectors, a **major sector** and a **minor sector**.

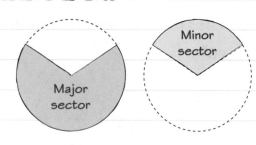

Major sector

Minor sector

You can find the area of a sector by working out what fraction it is of the whole circle.

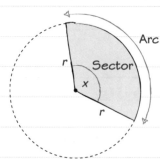

r

Arc

Sector

x

r

For a sector with angle x of a circle with radius r:

$$\text{Sector} = \frac{x}{360°} \text{ of the whole circle so}$$

$$\text{Area of sector} = \frac{x}{360°} \times \pi r^2$$

$$\text{Arc length} = \frac{x}{360°} \times 2\pi r$$

LEARN IT!

You can give answers in terms of π.
There is more about this on the previous page.

The diagram shows a minor sector of a circle of radius 13 cm.

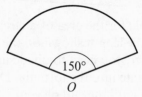

150°

O

Work out the perimeter of the sector. **(4 marks)**

$$\text{Arc length} = \frac{x}{360°} \times 2\pi r$$

$$= \frac{150°}{360°} \times 2\pi \times 13$$

$$= 34.033\,92...$$

$$\text{Perimeter} = \text{arc length} + \text{radius} + \text{radius}$$

$$= 34.033\,92... + 13 + 13$$

$$= 60 \text{ cm } (2 \text{ s.f.})$$

Don't round until your final answer. The radius is given correct to 2 significant figures so this is a good degree of accuracy.

Finding a missing angle

You can use the formulae for arc length or area to find a missing angle in a sector. Practise this method to help you tackle the hardest questions.

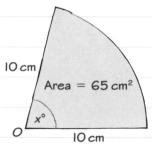

10 cm

Area = 65 cm²

x°

O

10 cm

$$\text{Area of sector} = \frac{x}{360} \times \pi r^2$$

$$65 = \frac{x}{360} \times \pi(10)^2$$

$$x = \frac{65 \times 360}{\pi(10)^2}$$

$$= 74.4845...$$

$$= 74.5° \text{ (to 3 s.f.)}$$

OAB is a sector of a circle, centre O.
Angle $AOB = 60°$.
$OA = OB = 12$ cm.
Work out the length of the arc AB.
Give your answer correct to 3 significant figures. **(3 marks)**

You need to learn the formula for arc length.

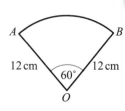

A   B

12 cm   12 cm

60°

O

Volumes of 3D shapes

If you need any of these formulae in your exam, they will be given to you with the question.

Cone

Volume of cone

$= \frac{1}{3} \times$ area of base $\times$ vertical height

$= \frac{1}{3} \pi r^2 h$

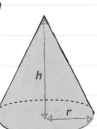

Sphere

Volume of sphere

$= \frac{4}{3} \pi r^3$

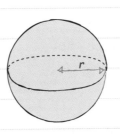

Pyramid

Volume of pyramid

$= \frac{1}{3} \times$ area of base $\times$ vertical height

$= \frac{1}{3} Ah$

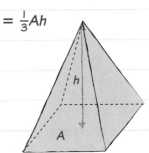

Worked example

The diagram shows a cone **A** and a cylinder **B**.
Show that the volume of **B** is 8 times the
volume of A.   **(4 marks)**

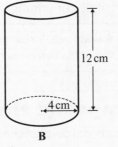

Volume of A $= \frac{1}{3}\pi r^2 h = \frac{1}{3}\pi \times 3^2 \times 8$
$= 24\pi$

Volume of B $= \pi r^2 h = \pi \times 4^2 \times 12$
$= 192\pi$

$8 \times 24\pi = 192\pi$ so the volume of B is
8 times the volume of A.

Problem solved!

You might have to **compare** two volumes or
areas in your exam. These questions might
involve:
- working out the ratio between two different
  areas or volumes
- finding an unknown quantity represented
  by a letter
- finding an expression for a length, area or
  volume in terms of an unknown.

In this question you need to know the ratio
between the two volumes. Calculate them
both, then write a short **conclusion**. Make sure
you show the calculation $8 \times 24\pi = 192\pi$ in
your conclusion. You can leave your working
in terms of π to make it easier. There is more
about this on page 83.

> You will need to use problem-solving skills
> throughout your exam – **be prepared!**

Now try this

The diagram shows a metal rivet made by
joining a hemisphere to a cylinder.

Work out the volume of metal used to
make the rivet. Give your answer correct to
3 significant figures.   **(4 marks)**

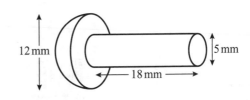

Worked solution
video

> You would be given the formula for the
> volume of a sphere with this question.

Surface area

Cone

The formula for the **curved surface area** of a cone will be given if you need it for a question.

Curved surface area of cone = $\pi r l$

> Be careful! This formula uses the slant height, l, of the cone.

To calculate the **total** surface area of the cone you need to add the area of the base. Surface area of cone = $\pi r^2 + \pi r l$

Sphere

The formula for the surface area of a sphere will be given if you need to use it.

Surface area of sphere = $4\pi r^2$

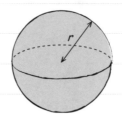

> For a reminder about areas of circles and surface areas of cylinders have a look at pages 104 and 106.

A hemisphere is half a sphere, so the area of the curved surface is $\frac{1}{2} \times 4\pi r^2$.

Worked example

The diagram shows a cone with vertical height 12 cm and base radius 5 cm. Work out the curved surface area of the cone.
(4 marks)

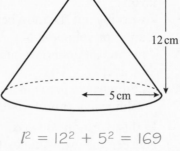

12 cm

5 cm

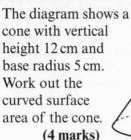

12 cm

l

5 cm

$l^2 = 12^2 + 5^2 = 169$

$l = 13\text{ cm}$

Curved surface area
= $\pi r l$
= $\pi \times 13 \times 5$
= $65\pi\text{ cm}^2$

Problem solved!

To work out the curved surface area you need to know the radius and the slant height. You are given the radius and the **vertical height**.
To calculate the slant height you need to use Pythagoras' theorem. Sketch the right-angled triangle containing the missing length.

> You will need to use problem-solving skills throughout your exam – **be prepared!**

Compound shapes

You can calculate the surface area of more complicated shapes by adding together the surface area of each part.

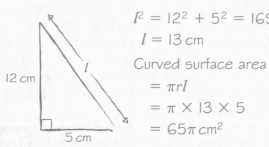

4 cm

6 cm

Surface area = $\pi(4)^2 + 2\pi(4)(6) + \frac{1}{2}[4\pi(4)^2]$
= $96\pi\text{ cm}^2$

Now try this

1 Work out the surface area of this prism.
(3 marks)

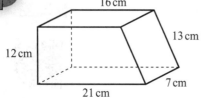

16 cm

13 cm

12 cm

21 cm

7 cm

Worked solution video

2 A solid object is formed by joining a hemisphere to a cylinder. Both the hemisphere and the cylinder have a diameter of 4.2 cm. The cylinder has a height of 5.6 cm. Work out the total surface area of the object. Give your answer to 3 significant figures.
(4 marks)

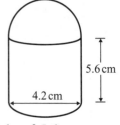

5.6 cm

4.2 cm

> Look at the blue box above for a hint.

Plans and elevations

Plans and elevations are 2D drawings of 3D shapes as seen from different directions.

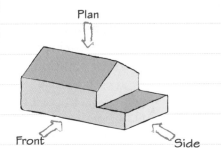

The **plan** is the view from above.

The **front elevation** is the view from the front.

The **side elevation** is the view from the side.

This line shows a change in depth.

You might have to deduce properties of a solid or sketch it from a plan and elevations.

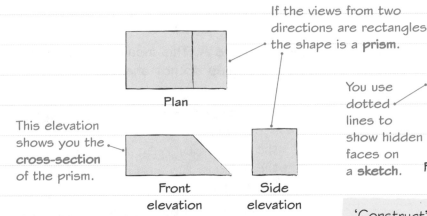

If the views from two directions are rectangles the shape is a **prism**.

This elevation shows you the **cross-section** of the prism.

You use dotted lines to show hidden faces on a **sketch**.

Worked example

The diagram shows a triangular prism. Accurately construct a plan and elevations for this prism.

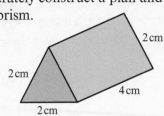

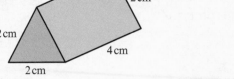

(4 marks)

Everything in red is part of the answer.

'Construct' means draw accurately using a pencil, ruler, compasses and protractor. Don't rub out any construction lines. Be careful with the side elevation. You need to use Pythagoras to work out the height.

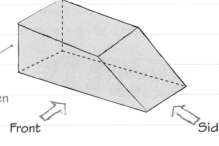

$$h^2 = \sqrt{2^2 - 1^2} = 1.7 \text{ (1 d.p.)}$$

Now try this

Here are the plan and elevations of a 3D shape.

| Front elevation | Side elevation | Plan |

For this 3D shape write down the number of
(a) faces   **(1 mark)**
(b) edges   **(1 mark)**
(c) vertices   **(1 mark)**

Sketch the 3D shape

Translations, reflections and rotations

You might have to describe these transformations in your exam. To describe a translation you need to give a vector. To describe a reflection you need to give the equation of the mirror line. To describe a rotation you need to give the direction, the angle and the centre of rotation.

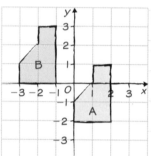

A to B: **Translation** by the vector $\begin{pmatrix} -3 \\ 2 \end{pmatrix}$

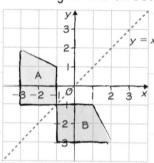

A to B: **Reflection** in the line $y = x$

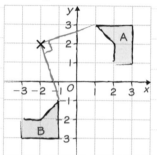

A to B: **Rotation** 90° clockwise about the point $(-2, 2)$

You can ask for tracing paper in an exam. This makes it easy to rotate shapes and check your answers.

For all three transformations, shape **B** is **congruent** to shape **A**. This means that they are exactly the same shape and size. Lengths of sides and angles do not change.

Worked example

The diagram shows two shapes **P** and **Q**.

Describe fully the single transformation which takes shape **P** to shape **Q**.   **(3 marks)**

Rotation 90° clockwise with centre (3, 1).

To **fully describe** a rotation you need to write down:
- the word 'rotation'
- the angle of turn and the direction
- the centre of rotation.

Be careful when the shapes are joined at one corner. This point is not necessarily the centre of rotation.

Check it!

You are allowed to ask for tracing paper in your exam. Trace shape **P** and put your pencil on your centre of rotation. Rotate the tracing paper to see if the shapes match up. ✓

Now try this

Triangles **A**, **B** and **C** are shown on the grid.

(a) Describe fully the **single** transformation that takes triangle **A** onto triangle **B**.   **(3 marks)**

(b) Describe fully the **single** transformation that takes triangle **A** onto triangle **C**.   **(1 mark)**

For part (a) you must write a **single** transformation. You will never have to write a combined transformation in your exam.

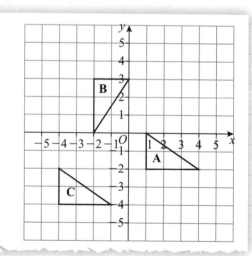

Enlargement

To describe an enlargement you need to give the scale factor and the centre of enlargement.

The **scale factor** of an enlargement tells you how much each length is multiplied by.

$$\text{Scale factor} = \frac{\text{enlarged length}}{\text{original length}}$$

Lines drawn through corresponding points on the object (**A**) and image (**B**) meet at the **centre of enlargement**.

When the scale factor is between 0 and 1, image **B** is **smaller** than object **A**.

When the scale factor is negative, image **B** is on the **other side** of the centre of enlargement and is upside down.

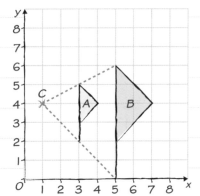

A to B: Each point on **B** is twice as far from **C** as the corresponding point on **A**.

Enlargement with scale factor 2, centre (1, 4).

For enlargements, angles in shapes do not change but lengths of sides do change.

Worked example

Triangle **A** is shown on the grid.

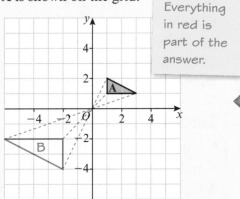

> Everything in red is part of the answer.

Enlarge triangle **A** by scale factor −2 with centre of enlargement O. **(3 marks)**

The scale factor is **negative** so the image will be on the **opposite** side of the centre of enlargement and will be **upside down**. Follow these steps to enlarge the triangle:

1. Draw lines from each vertex through the centre of enlargement.

2. Measure the distance from each point on triangle **A** to the centre of enlargement. The corresponding point on triangle **B** will be 2 times the distance from the centre of enlargement.

Check it!
Each length on triangle **B** should be 2 times the corresponding length on triangle **A**. ✓

Now try this

Triangle **A** is shown on the grid.

(a) Enlarge triangle **A** by scale factor $\frac{1}{2}$ with centre of enlargement (−3, 1).
Label the image **B**. **(2 marks)**

(b) Enlarge triangle **A** by scale factor −1$\frac{1}{2}$ with centre of enlargement (0, 1).
Label the image **C**. **(3 marks)**

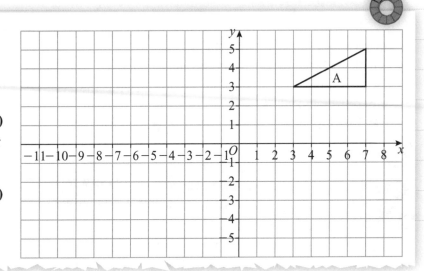

Combining transformations

You can describe two or more transformations using a single transformation.

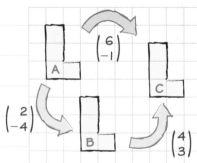

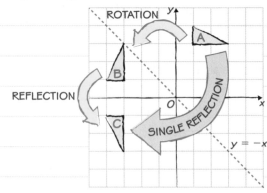

A to B to C: A translation $\begin{pmatrix} 2 \\ -4 \end{pmatrix}$ followed by a translation $\begin{pmatrix} 4 \\ 3 \end{pmatrix}$ is the same as a single translation $\begin{pmatrix} 6 \\ -1 \end{pmatrix}$.

A to B to C: A rotation 90° clockwise about O followed by a reflection in the x-axis is the same as a single reflection in the line $y = -x$.

Worked example

Triangle **A** is shown on the grid.

(a) Reflect triangle A in the y-axis. Label your new triangle **B**.
    (1 mark)

(b) Triangle **B** is reflected in the line $y = 1$ to give triangle **C**. Describe fully the **single** transformation which takes triangle **A** onto triangle **C**.   **(4 marks)**

Rotation 180° about the point (0, 1)

Everything in red is part of the answer.

To answer part (b) you need to draw
• the line $y = 1$
• the new triangle, **C**.

For a rotation of 180° you don't need to give a direction.

Check it!
You can ask for tracing paper in the exam.

Check a reflection by folding the tracing paper along the symmetry line. ✓

Describe fully ...

A translation: vector of translation.

A reflection: equation of mirror line.

A rotation: angle of turn, direction of turn and centre of rotation.

An enlargement: scale factor and centre of enlargement.

Now try this

Joelle reflects a shape in the line $x = 4$.
She then reflects the image in the line $y = 5$.

Describe fully a **single** transformation which has the same effect as Joelle's two transformations. You can use this graph paper to help you.   **(4 marks)**

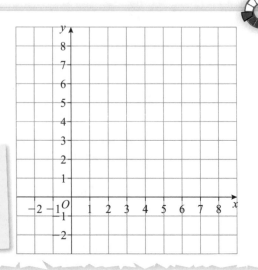

Worked solution video

Draw a triangle on the grid and apply the two transformations to help you see what is going on. You can use a second shape to check your answer. Make sure you write a **single** transformation in your answer.

Bearings

Bearings are measured **clockwise** from **north**.

Bearings always have **three figures**, so you need to add zeros if the angle is less than 100°. For instance, in this diagram the bearing of B from A is 048°.

You can measure a bearing bigger than 180° by measuring this angle and subtracting it from 360°.

The bearing of C from A is 360° − 109° = 251°

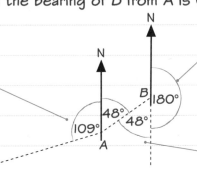

You can work out a reverse bearing by adding or subtracting 180°.

The bearing of A from B is 180° + 048° = 228°

These are alternate angles.

Worked example

The diagram shows the location of a sailing dinghy on a reservoir. The reservoir has a water inlet that is 700 m from the dinghy on a bearing of 230°.

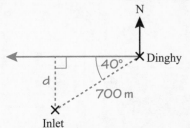

Everything in red is part of the answer.

(a) Write down the bearing of the dinghy from the water inlet. **(1 mark)**

230 − 180 = 50

Bearing of dinghy from inlet is 050°.

Boats are not allowed within 400 m of the water inlet. The dinghy sails due west.

(b) Show **by calculation** that the boat will not pass within 400 m of the water inlet. **(4 marks)**

$\sin x° = \dfrac{\text{opp}}{\text{hyp}}$

$\sin 40° = \dfrac{d}{700}$ so $d = 700 × \sin 40°$

$= 449.951... \text{ m}$

$d > 400$ m so the boat does not come within 400 m of the inlet.

Compass points

You need to know the compass points:

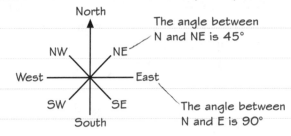

The angle between N and NE is 45°

The angle between N and E is 90°

Problem solved!

You might have to combine a bearings question with **trigonometry** calculations. You should always use a clear, well-labelled sketch to make sure you don't make any mistakes.

The shortest distance from a point to a line is a **perpendicular** line. Sketch this on the diagram and write down any lengths or angles you know. You can find the 40° angle by working out 270° − 230°.

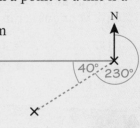

You will need to use problem-solving skills throughout your exam – **be prepared!**

Now try this

The diagram shows three orienteering markers, P, Q and R.

R is due east of Q. The bearing of Q from P is 068°. PQ = QR.

(a) Work out the bearing of P from Q. **(2 marks)**

(b) Work out the bearing of R from P. **(3 marks)**

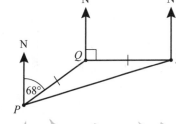

Worked solution video

Scale drawings and maps

This is a **scale drawing** of the Queen Mary II cruise ship.

Scale = 1 : 500

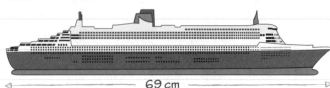

← 69 cm →

You can use the scale to work out the length of the actual ship.

69 × 500 = 34 500

The ship is 34 500 cm or 345 m long.

Map scales

Map scales can be written in different ways:

- 1 to 25 000
- 1 cm represents 25 000 cm
- 1 cm represents 250 m
- 4 cm represent 1 km

MAP
SCALE
1 : 25 000

Worked example

The diagram shows a scale drawing of a port and a lighthouse.

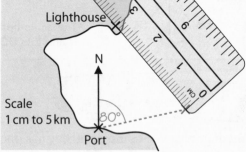

Lighthouse

N

Scale
1 cm to 5 km

80°

Port

A boat sails 12 km from the port in a straight line on a bearing of 080°.

How far away is the boat from the lighthouse? Give your answer in km.   **(3 mark)**

15 km

You need to use the scale drawing to answer the question. First, you need to mark the position of the boat accurately on the scale drawing.

Then work out how far the boat is from the port on the scale drawing.

	÷5	×12	
Map	1 cm	0.2 cm	2.4 cm
Real life	5 km	1 km	12 km
	÷5	×12	

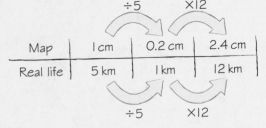

Now place the centre of your protractor on the port with the zero line pointing north. Put a dot at 80°. Line up your ruler between the port and the dot. Draw a cross 2.4 cm from the port.

Use a ruler to measure the distance from the lighthouse to the boat. 3 cm on the drawing represents 15 km in real life.

Now try this

1 A map has a scale of 1 : 40 000. A park is shown on the map as a rectangle measuring 5.8 cm by 4.4 cm.

   Calculate the area of the park in real life. Give your answer in km² to 3 significant figures.   **(3 marks)**

2 The map shows Hereford and Worcester. The scale of the map is 1 : 1 000 000. Work out the distance between Hereford and Worcester in kilometres.   **(3 marks)**

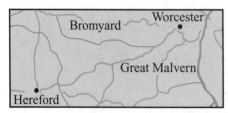

Bromyard   Worcester

Great Malvern

Hereford

You will need to measure the distance.

Constructions 1

You might be asked to construct a perpendicular line in any of these three ways.

Worked example

Use ruler and compasses to **construct** the perpendicular to the line segment AB that passes through point P.

(2 marks)

Everything in red is part of the answer.

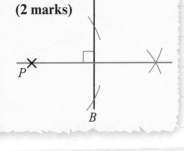

Use your compasses to mark two points on the line an equal distance from P. Keep the compasses the same and draw two arcs with their centres at these points.

Worked example

Use ruler and compasses to **construct** the perpendicular to the line segment AB that passes through point P.
(2 marks)

Use your compasses to mark two points an equal distance from P. Then widen your compasses and draw arcs with their centres at these two points.

Use your compasses to draw intersecting arcs with centres at A and B. Remember that to get full marks you have to be **accurate** and show **all** your construction lines.

Worked example

Use ruler and compasses to **construct** the perpendicular bisector of the line AB.
(2 marks)

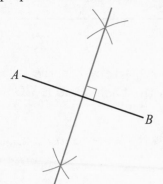

Everything in red is part of the answer.

Constructions checklist

Use good compasses with stiff arms. ✓
Use a sharp pencil. ✓
Use a transparent ruler. ✓
Mark any angles. ✓
Label any lengths. ✓
Show all construction lines. ✓

Now try this

Construct the perpendicular bisector of the line AB.   **(2 marks)**

Worked solution video

Remember to show **all** your construction marks.

Constructions 2

You need to know all of these constructions for your exam.

Use ruler and compasses to **construct** a triangle with sides of length 3 cm, 4 cm and 5.5 cm. **(2 marks)**

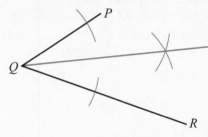

4 cm   3 cm
5.5 cm

Draw and label one side with a ruler. Then use your compasses to find the other vertex.

Use ruler and compasses to **construct** the bisector of angle *PQR*. **(2 marks)**

P
Q
R

Mark points on each arm an equal distance from *Q*. Then use arcs to find a third point an equal distance from these two points.

Use ruler and compasses to **construct** a 45° angle at *P*. **(2 marks)**

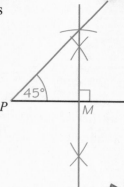

45°
P   *M*

Construct the perpendicular bisector of the line. Mark the midpoint *M*. Then set your compasses to the distance *PM*. Draw an arc on your bisector and join this point to *P* with a ruler.

Use ruler and compasses to **construct** a 60° angle at *P*. **(2 marks)**

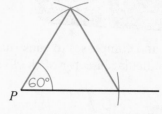

60°
P

Construct an equilateral triangle (all sides the same length). Each angle is 60°.

Everything in red is part of the answer.

1 Construct the bisector of the angle *ABC*. **(2 marks)**

2 Use ruler and compasses to construct an angle of 30°. **(3 marks)**

A
B   *C*

Worked solution video

You can construct a 30° angle by constructing a 60° angle then bisecting it.

Loci

A **locus** is a set of points which satisfy a condition. You can construct loci using ruler and compasses. A set of points can lie inside a **region** rather than on a line or curve.

The locus of points which are 7 cm from A is the circle, centre A.

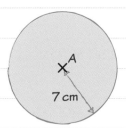

The region of points less than 7 cm from A lies inside this circle.

The locus of points which are the **same distance** from B as from C is the perpendicular bisector of BC.

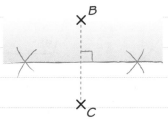

Points in the shaded region are closer to B than to C.

The locus of points which are 2 cm away from ST consists of two semicircles and two straight lines.

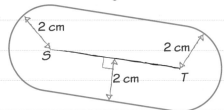

The shortest distance between a point and a line is a **perpendicular** from the point to the line.

Combining conditions

You can be asked to shade a region which satisfies more than one condition.

Here, the shaded region is more than 6 cm from point D **and** closer to line BC than to line AD.

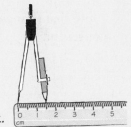

Worked example

This diagram is drawn accurately and shows part of a beach and the sea.
1 cm represents 20 m.

There is a lifeguard tower at point *P*. Public swimming is allowed in a region of the sea less than 30 m from the lifeguard tower. Shade this region on the diagram. **(2 marks)**

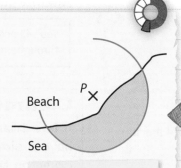

Everything in red is part of the answer.

I cm represents 20 m so 1.5 cm represents 30 m.

There is more about scale drawings on page 92.

Set your compasses to 1.5 cm. Set your compasses accurately by placing the point **on top** of your ruler at the 0 mark.

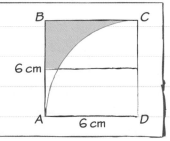

Now try this

A piece of card in the shape of an equilateral triangle *ABC* is placed on a horizontal straight line.

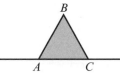

The card is first rotated 120° clockwise about *C*.
The card is then rotated 120° clockwise about *B*.
Draw the locus of the vertex *A*. **(3 marks)**

Start with your compasses point at *C* and your pencil at *A* and draw an arc. You need to work out where *B* touches the ground and move your compass point to there.

Congruent triangles

Two triangles are **congruent** if they have exactly the same shape and size.
To prove this you have to show that **one** of these four conditions is true.

1 **SSS** (three sides are equal)

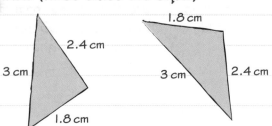

2.4 cm   1.8 cm
3 cm   3 cm   2.4 cm
1.8 cm

2 **AAS** (two angles and a corresponding side are equal)

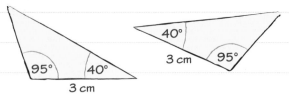
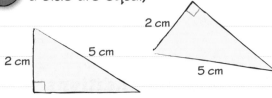

95°   40°   40°   95°
3 cm   3 cm

3 **SAS** (two sides and the included angle are equal)

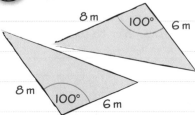
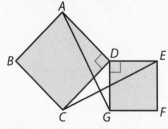

8 m   100°   6 m
8 m   100°   6 m

The angle must be **between** the two sides for SAS.

4 **RHS** (right angle, hypotenuse and a side are equal)

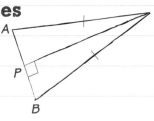

2 cm   5 cm   2 cm
5 cm

Problem solved!

You need to show that one of the conditions for congruence is true. You can only use the properties given in the question.

Always write down the condition for congruence.

> You will need to use problem-solving skills throughout your exam – **be prepared!**

Common sides

If two triangles have a side in **common** then those two sides are equal.

C
A
P
B

PC is common to triangles *APC* and *BPC*. Both triangles have a right angle and the same hypotenuse, so they satisfy RHS and are congruent.

Now try this

The diagram shows a kite *ABCD*.
AB = AD
BC = DC
Diagonals *AC* and *BD* intersect at *E*.
Prove that triangle *ABC* is congruent to triangle *ADC*.   **(3 marks)**

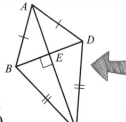

A
D
B   E
C

Remember to show **three** different sides or angles equal, and to write down which condition for congruence you are using: SSS, AAS, SAS or RHS

Worked solution video

Similar shapes 1

Shapes are **similar** if one shape is an enlargement of the other.

Similar triangles satisfy these three conditions:

 All three pairs of angles are equal.

 All three pairs of sides are in the same ratio.

 Two sides are in the same ratio and the included angle is equal.

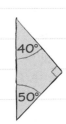

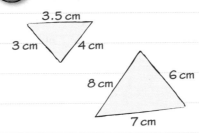

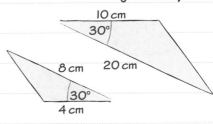

Worked example

XYZ and *ABC* are similar triangles.

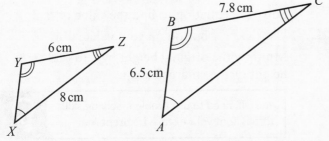

(a) Work out the length of *AC*. **(2 marks)**

$$\frac{AC}{XZ} = \frac{BC}{YZ}$$

$$\frac{AC}{8} = \frac{7.8}{6}$$

$$AC = \frac{7.8 \times 8}{6}$$

$$= 10.4 \text{ cm}$$

(b) Work out the length of *XY*. **(2 marks)**

$$\frac{XY}{AB} = \frac{YZ}{BC}$$

$$\frac{XY}{6.5} = \frac{6}{7.8}$$

$$XY = \frac{6 \times 6.5}{7.8}$$

$$= 5 \text{ cm}$$

Start with the unknown length on top of a fraction. Make sure you write your ratios in the correct order.

Similar shapes checklist

Use these facts to solve similar shapes problems:

Corresponding angles are equal.

Corresponding sides are in the same ratio.

Spotting similar triangles

Here are some similar triangles:

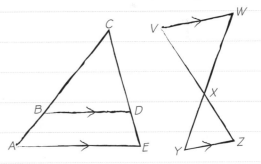

Triangle *ACE* is similar to triangle *BCD*

Triangle *VWX* is similar to triangle *ZYX*

Now try this

Triangles *ABC* and *PQR* are similar.

Angle *ACB* = angle *PRQ*

(a) Work out the size of angle *PRQ*. **(2 marks)**

(b) Work out the length of *PQ*. **(2 marks)**

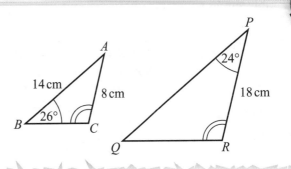

Similar shapes 2

The relationship between similar shapes is defined by a **scale factor**.
A and **B** are similar shapes. **B** is an enlargement of **A** with scale factor k.

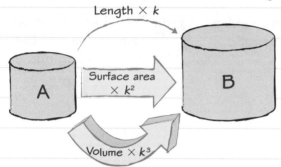

Length × k

Surface area × k^2

Volume × k^3

When a shape is enlarged by a linear scale factor k:

- Enlarged surface area
    = k^2 × original surface area
- Enlarged volume
    = k^3 × original volume
- Enlarged mass = k^3 × original mass

Worked example

These two glass prisms are similar in shape.

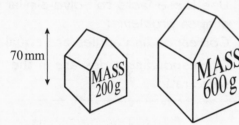

70 mm

MASS 200 g

MASS 600 g

The 200 g prism is 70 mm high.
Work out the height of the 600 g prism.

(3 marks)

$200 \times k^3 = 600$

$k^3 = \dfrac{600}{200} = 3$

$k = \sqrt[3]{3} = 1.4422\ldots$

Enlarged height = $70 \times 1.4422\ldots$
                      = 101 mm (3 s.f.)

Problem solved!

Always use k, k^2 or k^3 to write the relationship.
Enlarged mass = k^3 × original mass
Solve the equation to find the value of k.
Use the button on your calculator.
Multiply the original height by k to find the height of the enlarged shape.

You will need to use problem-solving skills throughout your exam – **be prepared!**

Comparing volumes

You can use k^3 to compare volume, mass or capacity.

$k = \dfrac{32}{16} = 2$

Volume of large bottle
= $1.2 \times k^3$
= 1.2×8
= 9.6 litres

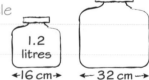

1.2 litres

←16 cm→   ← 32 cm →

Now try this

Here are three mathematically similar containers.
The table shows some information about these containers.

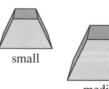

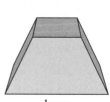

small

medium

large

	Height (cm)	Area of top of container (cm²)	Volume (cm³)
small	12	85	Y
medium	36	X	7830
large	W	2125	Z

Work out the missing values, W, X, Y and Z.

(6 marks)

The sine rule

The **sine rule** applies to any triangle. You don't need a right angle.

You label the angles of the triangle with capital letters and the sides with lower case letters. Each side has the same letter as its **opposite** angle.

$$\frac{a}{\sin A} = \frac{b}{\sin B} = \frac{c}{\sin C}$$

LEARN IT!

Use this version to find an unknown side.

$$\frac{\sin A}{a} = \frac{\sin B}{b} = \frac{\sin C}{c}$$

Use this version to find an unknown angle.

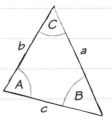

Worked example

Calculate the size of angle x.   **(3 marks)**

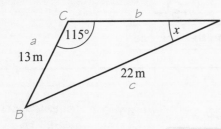

$$\frac{\sin A}{a} = \frac{\sin C}{c}$$

$$\frac{\sin x}{13} = \frac{\sin 115°}{22}$$

$$\sin x = \frac{13 \times \sin 115°}{22}$$

$$= 0.5355...$$

$$x = 32.4° \ (3 \text{ s.f.})$$

Everything in red is part of the answer.

Golden rule

To use the sine rule you need to know a side length and the **opposite** angle.

This is not a right-angled triangle so you can't use $S^O_H \ C^A_H \ T^O_A$. To find an angle use the 'upside down' version of the sine rule. You're not interested in side b or angle B so ignore this part of the rule.

Start by writing the value you want to find on top of the first fraction. Then substitute the other values you know and solve an equation to find x. Use the $\sin^{-1}$ function on your calculator.

Worked example

Calculate the length of AC.   **(3 marks)**

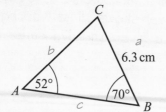

$$\frac{b}{\sin B} = \frac{a}{\sin A}$$

$$\frac{AC}{\sin 70°} = \frac{6.3}{\sin 52°}$$

$$AC = \frac{6.3 \times \sin 70°}{\sin 52°}$$

$$= 7.5126...$$

$$= 7.5 \text{ cm} \ (2 \text{ s.f.})$$

Everything in red is part of the answer.

You know a side length and the opposite angle so you can use the sine rule.

Check it!
The greater side length is opposite the greater angle. ✓

Now try this

In triangle ABC, $AB = 11$ cm, $AC = 13$ cm and angle $ABC = 57°$.

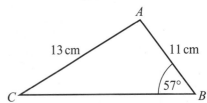

(a) Work out the size of angle ACB.   **(3 marks)**
(b) Work out the length of BC.   **(3 marks)**

The cosine rule

The **cosine rule** applies to any triangle. You don't need a right angle.

You usually use the cosine rule when you are given two sides and the included angle (SAS) or when you are given three sides and want to work out an angle (SSS).

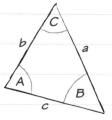

$$a^2 = b^2 + c^2 - 2bc \cos A$$

$$\cos A = \frac{b^2 + c^2 - a^2}{2bc}$$

Use this version to find an unknown side.

Use this version to find an unknown angle.

Which rule?

This chart shows you which rule to use when solving trigonometry problems in triangles:

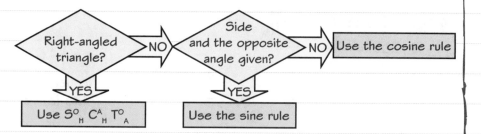

Worked example

PQRS is a trapezium. Work out the length of the diagonal *PR*. **(3 marks)**

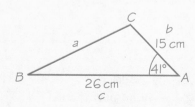

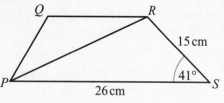

$a^2 = b^2 + c^2 - 2bc \cos A$

$PR^2 = 15^2 + 26^2 - 2 \times 15 \times 26 \times \cos 41°$

$ = 312.3265...$

$PR = 17.6727... = 17.7 \text{ cm (3 s.f.)}$

If you are given a more complicated diagram it is sometimes useful to sketch a triangle. Label your triangle with *a* as the missing side.

This is not a right-angled triangle so you can't use $S^O_H C^A_H T^O_A$. You know two sides and the included angle (SAS) so you can use the cosine rule.

Substitute the values you know into the formula. Work out the right-hand side using your calculator, but don't round your answer yet.

Use $\boxed{\sqrt{}}$ $\boxed{\text{Ans}}$ on your calculator to find the final answer.

Round to 3 significant figures or to the same degree of accuracy as the original measurements. Because this answer shows all its workings, you could give either 17.7 cm or 18 cm as the answer.

Now try this

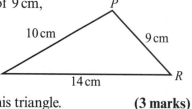

Triangle *PQR* has sides of 9 cm, 10 cm and 14 cm.

Work out the size of the smallest angle in this triangle. **(3 marks)**

Use the cosine rule if you are given **three sides** and you need to find an angle. You should use this version of the cosine rule:

$$\cos A = \frac{b^2 + c^2 - a^2}{2bc}$$

The smallest angle is always opposite the smallest side.

Triangles and segments

When you know the lengths of two sides and the angle **between them**, the area of any triangle can be found using this formula:

Area = $\frac{1}{2}$ ab sin C

You can use this formula for **any** triangle. You don't need to have a right angle.

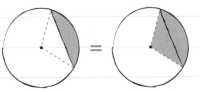

LEARN IT!

Areas of segments

A chord divides a circle into two **segments**.

Area of minor segment = Area of whole sector − Area of triangle

---

Worked example

The diagram shows a sector of a circle with centre O.

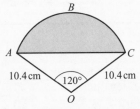

Work out the area of the shaded segment ABC. Give your answer correct to 3 significant figures. **(5 marks)**

Whole sector $OABC$:

Area = $\frac{120}{360}$ × π × 10.4²

     = 113.2648... cm²

Triangle OAC:

Area = $\frac{1}{2}$ × 10.4 × 10.4 × sin 120°

     = 46.8346... cm²

Shaded segment ABC:

Area = 113.2648... − 46.8346...

     = 66.4302...

     = 66.4 cm² (to 3 s.f.)

If you are aiming for a top grade you need to be able to calculate the area of a sector and a triangle.

To get full marks you need to keep track of your working. Make sure you write down exactly what you are calculating at each step.

Remember that 10.4 cm is the length of one side of the triangle **and** the radius of the circle.

Make sure you don't round too soon. Write down all the figures from your calculator display at each step. Only round your **final answer** to 3 significant figures.

Which formula?

If you know the base and the vertical height:

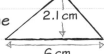

Area = $\frac{1}{2}$ × base × vertical height

     = $\frac{1}{2}$ × 6 × 2.1

     = 6.3 cm²

If you know two sides and the included angle:

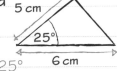

Area = $\frac{1}{2}$ ab sin C

     = $\frac{1}{2}$ × 5 × 6 × sin 25°

     = 6.3... cm²

Now try this

The diagram shows a circle, radius 4 cm, with angle $AOB = 135°$.

Using the fact that sin 135° = sin 45°, show clearly that the shaded area is equal to $6\pi - 4\sqrt{2}$ cm². **(5 marks)**

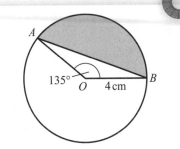

This could be a **non-calculator** question so you need to know the exact value of sin 45°. Have a look at page 79 for a reminder.

Pythagoras in 3D

To tackle the most demanding questions, you need to be able to use Pythagoras' theorem in 3D shapes.

You can use Pythagoras' theorem to find the length of the longest diagonal in a cuboid.

You can also use Pythagoras to find missing lengths in pyramids and cones.

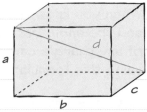

$$a^2 + b^2 + c^2 = d^2$$

Why does it work?

You can use 2D Pythagoras twice to show why the formula for 3D Pythagoras works.

$$x^2 = b^2 + c^2$$

$$d^2 = a^2 + x^2$$
$$= a^2 + b^2 + c^2$$

Worked example

The diagram shows a cuboid. Work out the length of *PQ*.
(3 marks)

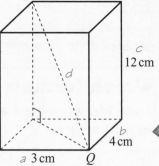

$$d^2 = a^2 + b^2 + c^2$$
$$PQ^2 = 3^2 + 4^2 + 12^2$$
$$\quad = 169$$
$$PQ = \sqrt{169} = 13$$
So *PQ* is 13 cm.

Everything in red is part of the answer.

Write out the formula for Pythagoras in 3D. Label the sides of the cuboid *a*, *b* and *c*, and label the long diagonal *d*.

You could also answer this question by sketching two right-angled triangles and using 2D Pythagoras.

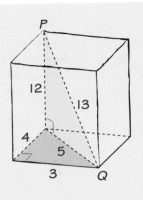

Check it!
The diagonal must be longer than any of the other three lengths.

13 cm looks about right. ✓

Now try this

The diagram shows Bridget's new sewing box and a knitting needle.

Will the knitting needle fit inside the box?

You must show all of your working.   **(3 marks)**

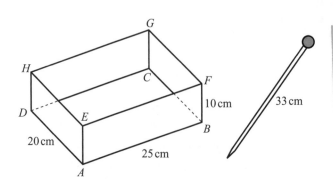

Worked solution video

Trigonometry in 3D

You can use $S^O_H C^A_H T^O_A$ to find the angle between a **line** and a **plane**.

You might need to combine trigonometry and Pythagoras' theorem when you are solving 3D problems.

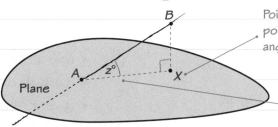

Point X is directly below point B, so ABX is a right-angled triangle.

Angle z is the angle between the line and the plane.

Worked example

The diagram shows a triangular prism.

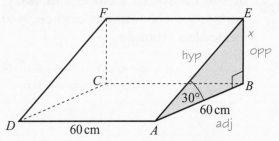

Calculate the angle between the line DE and the base of the prism. **(6 marks)**

△ABE: you know one angle and the adjacent side. You are looking for the opposite side, so use T^O_A.

△DAB: you know two sides so you can use Pythagoras' theorem.

△DEB: you know the opposite and adjacent sides so use T^O_A. Use $\tan^{-1}$ to find the value of z.

Do **not** round any of your answers until the end — write down at least six figures from each calculator display.

$\tan 30° = \dfrac{x}{60}$

$x = 60 \times \tan 30°$

$\quad = 34.6410... \text{ cm}$

Everything in red is part of the answer.

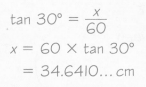

$y^2 = 60^2 + 60^2$

$\quad = 7200$

$y = \sqrt{7200} = 84.8528... \text{ cm}$

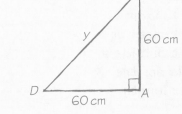

$\tan z° = \dfrac{34.6410...}{84.8528...} = 0.40824...$

$z° = \tan^{-1} 0.408\,24... = 22.2076...$

$\quad = 22.2°$ (to 3 s.f.)

Now try this

The diagram shows a triangular-based pyramid.

OAC and OCB are right-angled triangles.

Work out the size of angle BOC.

Give your answer correct to 1 decimal place. **(6 marks)**

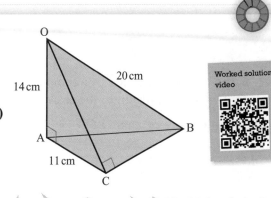

Worked solution video

Start by sketching triangle OAC and working out the length of OC.

Circle facts

You need to know the names of the different parts of a circle.

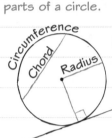

Diameter = radius × 2

Tangent

The other parts of a circle are shown on pages 83 and 84.

When you are solving circle problems:
- correctly identify the angle to be found
- use all the information given in the question
- mark all calculated angles on the diagram
- give a reason for each step of your working.

You might need to use angle facts about triangles, quadrilaterals and parallel lines in circle questions. There is a list of angle facts on page 74.

Key circle facts

1 The angle between a radius and a tangent is 90°.

2 Two tangents which meet at a point outside a circle are the same length.

3 A triangle which has one vertex at the centre of a circle and two vertices on the circumference is an **isosceles triangle**.

Each short side of the triangle is a radius, so they are the same length.

Remember that the base angles of an isosceles triangle are equal.

Worked example

A and B are points on the circumference of a circle centre O. AC and BC are both tangents to the circle. Angle BCA = 42°. Work out the size of the angle marked x. **(3 marks)**

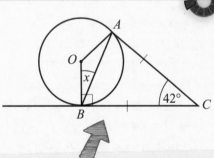

$AC = BC$ (tangents from a point outside a circle are the same length)

$\angle ABC = \dfrac{180° - 42°}{2} = 69°$
(base angles in an isosceles △ are equal, and angles in a △ add up to 180°)

$x + 69° = 90°$ (angle between a tangent and a radius = 90°)

$x = 21°$

AC = BC, so mark these lines with a dash. Make sure you also write down the circle fact you are using. To write a really good answer you have to give a reason for each step of your working.

Now try this

A, B and C are points on the circumference of a circle with centre O. CD is a tangent to the circle. Angle AOB = 53°

Work out the size of angle BCD.   **(3 marks)**

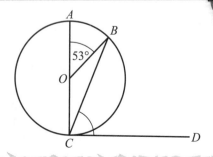

Make sure you write down every step of your working.

Circle theorems

You need to **learn** these six circle theorems and know how to apply them to solve problems.

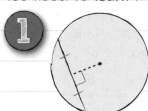 The perpendicular from a chord to the centre of the circle bisects the chord.

LEARN IT!

 Opposite angles of a cyclic quadrilateral add up to 180°.

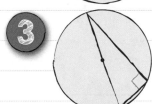 The angle in a semicircle is 90°.

 Angles in the same segment are equal.

 The angle at the centre of the circle is twice the angle on the circumference.

The angle between a tangent and a chord is equal to the angle in the alternate segment.

This is called the **alternate segment theorem**.

Turn to page 109 to have a go at **proving** this circle theorem.

Worked example

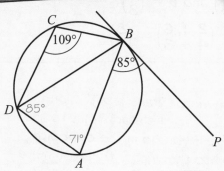

Everything in red is part of the answer.

Angle $ABP = 85°$ and angle $BCD = 109°$
Calculate the size of angle ABD.
Give reasons for each step of your working.   **(4 marks)**

$\angle ADB = 85°$ (alternate segment theorem)

$180 - 109 = 71$

$\angle DAB = 71°$ (opposite angles in a cyclic quadrilateral add up to 180°)

$180 - 85 - 71 = 24$

$\angle ABD = 24°$ (angles in a triangle add up to 180°)

Circle theorem top tips

☑ If you can spot a tangent and a chord in your circle then you might be able to use the **alternate segment theorem**.

☑ **Write** the unknown angles on your diagram as you go.

☑ Give **reasons** for each step of your working.

You might need to use other angle facts in a circle theorem question. Look at page 74 for a reminder.

Now try this

ABCD is a cyclic quadrilateral.
PBQ is a tangent to the circle at *B*.

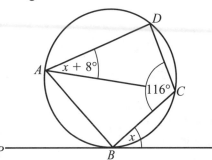

Work out the value of *x*.   **(4 marks)**

Vectors

A vector has a **magnitude** (or size) and a **direction**.

This vector can be written as **a**, $\overrightarrow{AB}$ or $\binom{2}{5}$.

You can multiply a vector by a number. The new vector has a different length but the same direction.

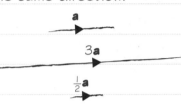

If **b** is a vector then −**b** is a vector with the same length but opposite direction.

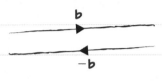

Worked example

In the diagram, *OADB* and *ACED* are two identical parallelograms.

a is the vector $\overrightarrow{OA}$
b is the vector $\overrightarrow{OB}$

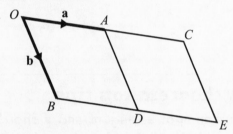

Find the following vectors in terms of **a** and **b**.

(a) $\overrightarrow{OD}$ **(1 mark)**

$\overrightarrow{OD} = \overrightarrow{OA} + \overrightarrow{AD}$

$= a + b$

(b) $\overrightarrow{EB}$ **(1 mark)**

$\overrightarrow{EB} = \overrightarrow{ED} + \overrightarrow{DB}$

$= -a + -a = -2a$

(c) $\overrightarrow{BC}$ **(1 mark)**

$\overrightarrow{BC} = \overrightarrow{BD} + \overrightarrow{DE} + \overrightarrow{EC}$

$= a + a + -b = 2a - b$

Adding vectors

You can add vectors using the **triangle law**. You trace a path along the added vectors to find the new vector.

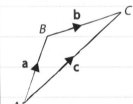

$a + b = c$

c is the resultant vector of **a** and **b**.

If $a = \binom{2}{4}$ and $b = \binom{6}{3}$

then $c = \binom{2+6}{4+3} = \binom{8}{7}$

For each vector, trace a path along the shape from the start point to the end point. If you go in the **opposite** direction to the vector then you need to **subtract**.
$\overrightarrow{AD} = \overrightarrow{OB}$ because they are **parallel**.

Simplify your vectors as much as possible.

If you multiply a column vector by a number you have to multiply **both parts**.

$2 \times \binom{p}{q} = \binom{2p}{2q}$

To **add** column vectors you add the top numbers and add the bottom numbers:

$\binom{c}{d} + \binom{e}{f} = \binom{c+e}{d+f}$

Now try this

The vectors **a** and **b** are defined as

$a = \binom{3}{5}$ $b = \binom{2}{-9}$

Write the following as column vectors.

(a) 2**a** **(1 mark)**

(b) **a** + **b** **(1 mark)**

(c) **b** − 3**a** **(3 marks)**

Vector proof

Parallel vectors

If one vector can be written as a **multiple** of the other then the vectors are **parallel**.

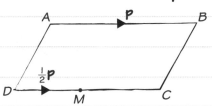

In this parallelogram M is the midpoint of DC.
AB is parallel to DM so $\overrightarrow{DM} = \frac{1}{2}\overrightarrow{AB}$

Remember that AB means the line segment AB (or the length of the line segment AB). $\overrightarrow{AB}$ means the vector which takes you from A to B.

Collinear points

If three points lie on the **same straight line** then they are collinear. Here are three points:

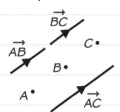

If any **two** of the vectors $\overrightarrow{AB}$, $\overrightarrow{BC}$ or $\overrightarrow{AC}$ are parallel, then the three points must be collinear.

Worked example

In the diagram, OPQ is a triangle. Point R lies on the line PQ such that $PR:RQ = 1:2$
Point S lies on the line through OR such that $OR:OS = 1:3$

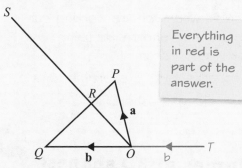

> Everything in red is part of the answer.

(a) Show that $\overrightarrow{OS} = 2\mathbf{a} + \mathbf{b}$ **(3 marks)**

$\overrightarrow{PQ} = -\mathbf{a} + \mathbf{b}$, so $\overrightarrow{PR} = \frac{1}{3}(-\mathbf{a} + \mathbf{b})$
$\overrightarrow{OR} = \mathbf{a} + \frac{1}{3}(-\mathbf{a} + \mathbf{b})$
$\quad = \frac{1}{3}(2\mathbf{a} + \mathbf{b})$
$\overrightarrow{OS} = 3\overrightarrow{OR}$
$\quad = 2\mathbf{a} + \mathbf{b}$

(b) Point T is added to the diagram such that $\overrightarrow{TO} = \mathbf{b}$. Prove that points T, P and S lie on the same straight line. **(3 marks)**

$\overrightarrow{TP} = \mathbf{a} + \mathbf{b}$
$\overrightarrow{TS} = \mathbf{b} + 2\mathbf{a} + \mathbf{b}$
$\quad = 2\mathbf{a} + 2\mathbf{b}$
$\quad = 2(\mathbf{a} + \mathbf{b})$
$\overrightarrow{TS} = 2\overrightarrow{TP}$ so $\overrightarrow{TS}$ and $\overrightarrow{TP}$ are parallel. Both vectors pass through T so they lie on the same straight line.

Problem solved!

You might be given information about the lengths of lines as ratios.

$PR:RQ = 1:2$ There are $2 + 1 = 3$ parts in this ratio. This means that R is $\frac{1}{3}$ of the way along PQ so $\overrightarrow{PR} = \frac{1}{3}\overrightarrow{PQ}$

$OR:OS = 1:3$ This means that $\overrightarrow{OS} = 3\overrightarrow{OR}$

To show that T, P and S lie on the same straight line you need to show that **two** of the vectors $\overrightarrow{TP}$, $\overrightarrow{TS}$ and $\overrightarrow{PS}$ are parallel.

> You will need to use problem-solving skills throughout your exam – **be prepared!**

Now try this

In triangle OAB, M is the midpoint of OA.
$OH = HJ = JK = KB$
S and T divide the line AB into three equal segments.
$\overrightarrow{OM} = \mathbf{a}$ and $\overrightarrow{OH} = \mathbf{b}$

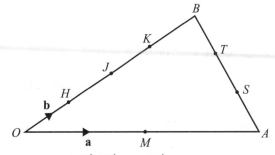

(a) Prove that $\overrightarrow{HA}$, $\overrightarrow{JS}$ and $\overrightarrow{KT}$ are all parallel.
(6 marks)

(b) State the ratio $KT:JS:HA$ **(1 mark)**

Had a look ☐ Nearly there ☐ Nailed it! ☐

Problem-solving practice 1

Throughout your Higher GCSE exam you will need to **problem solve**, **reason**, **interpret** and **communicate** mathematically. If you come across a tricky or unfamiliar question in your exam you can try some of these strategies:

- ✓ Sketch a diagram to see what is going on.
- ✓ Try the problem with smaller or easier numbers.
- ✓ Plan your strategy before you start.
- ✓ Write down any formulae you might be able to use.
- ✓ Use x or n to represent an unknown value.

1 A ladder is 6 m long.

The ladder is placed on horizontal ground, resting against a vertical wall.

The instructions for using the ladder say that the bottom of the ladder must not be closer than 1.5 m to the bottom of the wall.

How far up the wall can the ladder reach if the instructions are followed?　**(3 marks)**

Pythagoras' theorem page 76

You should definitely draw a sketch to show the information in the question.

TOP TIP

Be careful when you are working out the length of a **short** side using Pythagoras' theorem.

Remember: $\text{short}^2 + \text{short}^2 = \text{long}^2$

$\text{short}^2 = \text{long}^2 - \text{short}^2$

2 The diagram shows four metal ball-bearings of diameter 2.7 cm packed inside the smallest possible box in the shape of a cuboid.

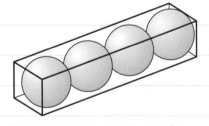

Worked solution video

Volume of a sphere = $\frac{4}{3}\pi r^3$

Work out the volume of empty space inside the box. Give your answer correct to 3 significant figures.　**(4 marks)**

Volumes of 3D shapes page 85

You need to work out the dimensions of the box. You can use the diagram to help – once you have worked out each dimension, write it on the diagram.

Don't round any values until your final answer. You can use the ANS button on your calculator to enter the exact answer to a previous calculation. But don't write 'ANS' in your working. It's better to write down the actual number with at least 4 decimal places.

TOP TIP

If you need to use the formula for the volume of a sphere you will be given it in the question.

Problem-solving practice 2

3

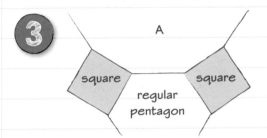

The diagram shows 2 squares, part of a regular pentagon and part of a regular *n*-sided polygon, A.

Calculate the value of *n*. Show your working clearly. **(5 marks)**

Angles in polygons page 75

Follow these steps:

1. Work out the interior angles of a regular pentagon and a square.
2. Use the fact that the angles around a point add up to 360° to find the interior angle of A.
3. Subtract this from 180° to find the exterior angle of A.
4. Divide 360° by this to work out *n*.

(**TOP TIP**)

If there are a lot of steps in a question it's a good idea to **plan** your answer before you start.

4

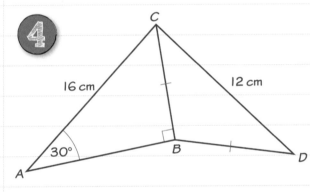

$AC = 16\,cm$ $CD = 12\,cm$ $BC = BD$

Angle $ABC = 90°$

Angle $CAB = 30°$

Work out the area of triangle BCD.

(6 marks)

The cosine rule page 100
Triangles and segments page 101

You need more information about triangle BCD before you can calculate its area.

Triangle ABC is right angled so you can use $S^O_H C^A_H T^O_A$ to work out the length of BC. Then use the cosine rule to work out the size of angle BCD. Finally, you can use the formula

$A = \frac{1}{2}ab \sin C$ to work out the area of triangle BCD.

(**TOP TIP**)

Unless it says so in the question, diagrams are **not drawn accurately**, so you can't measure lengths or angles.

5

The diagram shows a circle with centre O.

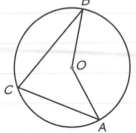

Prove that angle BOA is twice angle BCA. **(4 marks)**

Solving angle problems page 74
Circle theorems page 105

If you have to **prove** a fact about circles you should use only basic angle facts about triangles, straight lines and parallel lines. This diagram shows you how to get started.

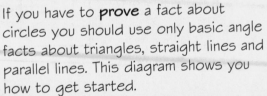

OA, OB and OC are all radii, so they are the same length. Use the fact that OCB and OCA are both **isosceles** triangles to prove that $q = 2p$ and $y = 2x$.

(**TOP TIP**)

It's OK to draw extra lines on the diagram and label extra angles. Your diagram is part of your working, so be neat and clear.

Mean, median and mode

You can analyse data by calculating statistics like the **mean**, **median** and **mode**.

Mean

Add up all the values

↓

Divide by the total number of values

↓

Do not round your answer

Median

Write the values in order of size, smallest first

↓

Count the number of values

← Odd number of values

The median is the middle value

→ Even number of values

The median is half-way between the two middle values

Mode

Look for the most common value

← One value with highest frequency

This value is the mode

↓ More than one value with highest frequency

These values are all modes

→ All values have same frequency

There is no mode

Worked example

Kayla has 8 numbered cards.

$\boxed{1}\boxed{2}\boxed{3}\boxed{4}\boxed{5}\boxed{6}\boxed{7}\boxed{8}$

She removes two cards. The mean value of the remaining cards is 4. Which two cards could Kayla have removed? Give **one** possible answer.

(4 marks)

You can work out the sum of the 6 remaining cards using this formula:

Sum of values = mean × number of values

Subtract this sum from the sum of all 8 cards. This tells you the sum of the 2 cards Kayla removed. The removed cards were either 5 and 7 or 8 and 4.

Check it!

Work out the mean of the remaining 6 cards.

$6 \times 4 = 24$

$1 + 2 + 3 + 4 + 5 + 6 + 7 + 8 = 36$

$36 - 24 = 12$

The removed cards add up to 12 so Kayla could have removed 7 and 5

Check:

$$\frac{1 + 2 + 3 + 4 + 6 + 8}{6} = 4 \checkmark$$

Which average works best?

	👍	👎
Mean	Uses all the data	Affected by extreme values
Median	Not affected by extreme values	Value may not exist
Mode	Suitable for data that can be described in words	Not always near the middle of the data

---

Now try this

Make sure you check your answer by calculating the new mean and median. Remember that the median is not affected by extreme values.

Joe scored these marks out of 20 in **six** maths tests.

    11   9   5   13   15   12

How many marks must he score in the next test so that his mean mark and his median mark are the same?

(3 marks)

Frequency table averages

This page shows you how to find averages from data given in frequency tables. Have a look at pages 118 and 121 to revise finding averages from graphs.

This frequency table shows the numbers of pets owned by the students in a class.

The mode is 1. This value has the highest frequency.

Number of pets (x)	Frequency (f)	Frequency × number of pets (f × x)
0	12	12 × 0 = 0
1	18	18 × 1 = 18
2	5	5 × 2 = 10
3	2	2 × 3 = 6
Total	37	34

To calculate the mean you need to add a column for 'f × x'.

There are 37 values so the median is the $\frac{37+1}{2} = $ 19th value.
The first 12 values are all 0. The next 18 values are 1. So the median is 1.

The total in the 'f × x' column represents the total number of pets owned by the class.

$$\text{Mean} = \frac{\text{total number of pets}}{\text{total frequency}} = \frac{34}{37} = 0.92 \text{ (to 2 d.p.)}$$

Worked example

Maisie recorded the times, in minutes, taken by 150 students to travel to school.

The table shows her results.

Time (t minutes)	Frequency (f)	Midpoint (x)	f × x
0 ≤ t < 20	65	10	65 × 10 = 650
20 ≤ t < 30	40	25	40 × 25 = 1000
30 ≤ t < 40	39	35	39 × 35 = 1365
40 ≤ t < 60	6	50	6 × 50 = 300

Total frequency = 150   Total of f × x = 3315

Everything in red is part of the answer.

(a) Work out an estimate for the mean number of minutes that the students took to travel to school.
(4 marks)

$$\frac{3315}{150} = 22.1$$

(b) Explain why your answer to part (a) is an estimate.
(1 mark)

Because you don't know the exact data values.

Add extra columns to the table for 'Midpoint (x)' and 'Midpoint × frequency (f × x)'.

 Estimate of mean = $\frac{\text{Total of } fx \text{ column}}{\text{Total frequency}}$

Your answer does not have to be a whole number. Write down all the digits from your calculator display then round **if necessary**.

Now try this

Emma recorded the reaction times of the members of her class using a computer program. The table shows her results.

Work out an estimate for the mean reaction time. **(4 marks)**

Reaction time (t seconds)	Frequency
0 ≤ t < 0.1	2
0.1 ≤ t < 0.2	9
0.2 ≤ t < 0.3	16
0.3 ≤ t < 0.4	5

Worked solution video

Interquartile range

Range and interquartile range are measures of spread. They tell you how spread out data is.

Quartiles divide a data set into four equal parts.

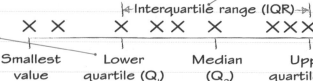

Half of the values lie between the lower quartile and the upper quartile.

$Q_1 = \dfrac{n+1}{4}$th value, where n = number of data values

Interquartile range (IQR)

× ×   × × × ×   × × × ×   ×

Smallest value   Lower quartile (Q_1)   Median (Q_2)   Upper quartile (Q_3)   Largest value

$Q_3 = \dfrac{3(n+1)}{4}$th value

Range = largest value − smallest value

Interquartile range (IQR) = upper quartile (Q_3) − lower quartile (Q_1)

Worked example

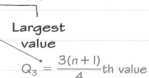

Alison recorded the heights, in cm, of some tree saplings. She put the heights in order.

21   23   23   25   26   26   31   32
33   35   36   40   40   41   42

Work out the interquartile range of Alison's data.

(3 marks)

$n = 15$

$\dfrac{n+1}{4} = \dfrac{15+1}{4} = 4$

$Q_1 = $ 4th value $= 25$ cm

$\dfrac{3(n+1)}{4} = \dfrac{3(15+1)}{4} = 12$

$Q_3 = $ 12th value $= 40$ cm

$IQR = Q_3 - Q_1 = 40 - 25 = 15$ cm

To work out the interquartile range, you need to know the lower quartile and the upper quartile.

1. Count the total number of values, n.

2. Check that the data is arranged in order of size.

3. Find the $\dfrac{n+1}{4}$th data value.
   This is the lower quartile (Q_1).

4. Find the $\dfrac{3(n+1)}{4}$th data value.
   This is the upper quartile (Q_3).

5. Subtract the lower quartile from the upper quartile to find the interquartile range.

Golden rule

Always arrange the data in order of size before calculating the median or quartiles.

If the data is given in an ordered **stem-and-leaf diagram** then it is already in order of size.

This stem-and-leaf diagram shows the costs, in £, of some DVDs. There are 11 pieces of data in this stem-and-leaf diagram.

0|9 represents £9

This is the stem.

```
0 | 7 9 9
1 | 0 0 2 3 5 7
2 | 0 5
```

Key: 1 | 5 = £15

There are 11 pieces of data, so the median is the 6th value. The median is £12.

Now try this

Worked solution video

Henry measures the heights of some plant seedlings for an experiment. This stem-and-leaf diagram shows his results.

(a) How many plant seedlings did Henry measure? **(1 mark)**

(b) Work out the median height. **(2 marks)**

(c) Work out the interquartile range of this data. **(3 marks)**

```
4 | 0 2 4 4 5 9
5 | 1 3 3 4 7 8 9
6 | 2 2 5 6 8
7 | 1
```

Key: 5 | 3 means 5.3 cm

Line graphs

A **time series** graph is a line graph that shows how a variable changes over a period of time. This time series graph shows median incomes in the UK from 1980 to 2010.

The graph should have a title. ———————→ UK median incomes

The **vertical axis** shows the variable which changes over time.

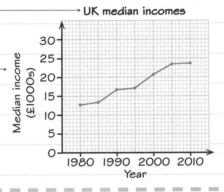

The graph shows an **upward trend**. This means that in general, as time passes, the median income **increases**.

The **horizontal axis** always shows units of time.

Vertical line graphs

A vertical line graph is sometimes called a **bar-line graph**. It works in the same way as a **bar chart** but the bars are straight, vertical lines.

Problem solved!

This is like finding the mean from this frequency table:

Number of pairs of trainers	0	1	2	3	4	5	6
Frequency	0	4	7	6	2	0	1

Work out the total number of pairs of trainers owned, then divide by the total number of members of the class.

You can revise how to find the mean from a frequency table on page 111.

You will need to use problem-solving skills throughout your exam – **be prepared!**

Worked example

This vertical line graph shows the number of pairs of trainers owned by the members of a class.

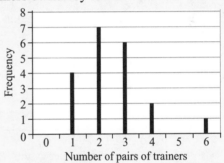

What was the mean number of pairs of trainers owned? **(2 marks)**

Total number of pairs of trainers owned
0 × 0 + 4 × 1 + 7 × 2 + 6 × 3 + 2 × 4 + 0 × 5 + 1 × 6 = 50
Total number of members of class
0 + 4 + 7 + 6 + 2 + 0 + 1 = 20
50 ÷ 20 = 2.5
The mean is 2.5 pairs of trainers.

Now try this

The table on the right gives information about the percentage of students who passed a professional exam each year from 2007 to 2013.

(a) Draw a time series graph to represent this data. **(3 marks)**

(b) Describe the trend. **(1 mark)**

Year	2007	2008	2009	2010	2011	2012	2013
Percentage (%)	58	60	53	55	63	65	73

You can use a 'break' in your vertical axis.

Scatter graphs

The points on a scatter graph aren't always scattered. If the points are almost on a straight line then the scatter graph shows **correlation**. The better the straight line, the stronger the correlation.

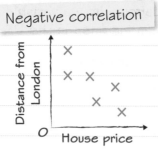

Negative correlation

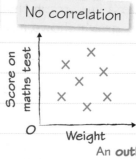

No correlation

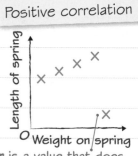

Positive correlation

An **outlier** is a value that does not fit the pattern of the data.

Worked example

This scatter graph shows the engine capacity of some cars and the distance they will travel on one gallon of petrol.

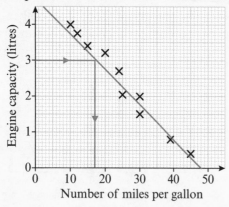

(a) Estimate how far a car with a 3-litre engine will travel on 1 gallon of petrol. **(2 marks)**

17 miles

(b) Comment on the reliability of your estimate. **(1 mark)**

3 litres is within the range of the data values (interpolation) and the correlation is strong, so the estimate is reliable.

Cause and effect

Watch out! Correlation doesn't always mean that two variables are related.

Bottled water doesn't cause bee stings but, when the weather is hotter, bottled water sales and bee stings both increase.

Use a **line of best fit** to estimate. Draw a straight line as close to as many of the points on the graph as possible.

Estimating and predicting

✓ If you are predicting a value that is **within** the range of the data your prediction will be **more accurate**.
This is called **interpolation**.

✗ If you are predicting a value that falls **outside** the range of the data your prediction will be **less accurate**.
This is called **extrapolation**.

Now try this

This scatter graph shows the daily hours of sunshine and the daily maximum temperature at 10 seaside resorts in England one day last summer.

(a) Another resort had 10 hours of sunshine each day. Use a line of best fit to predict the maximum temperature at this resort. **(2 marks)**

(b) Comment on the reliability of your estimate. **(1 mark)**

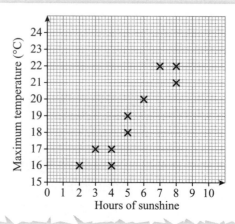

Sampling

In statistics, a **population** is a group of people you are interested in. A **sample** is a smaller group chosen from a larger population. You can use data from the sample to make **predictions** about the whole population.

Advantages of sampling

 It is **cheaper** to survey a sample than a whole population.

 It is **quicker** to collect data from a sample.

 It is **easier** to analyse data from a sample and calculate statistics.

Random sample

In a random sample, every member of the population has an equal chance of being included in the sample. Here are two ways of selecting a random sample:

 Put the names of every member of the population in a hat and select your sample at random.

 Assign a number to every member of the population and choose random numbers using a computer program or calculator.

Problem solved!

In part (a) you are using the sample to estimate the mean for the whole population. You calculate the mean of the sample data in the normal way. But this is only an **estimate** of the mean of the whole population. The accuracy of the estimate depends on whether the sample is **biased** or not. There are two reasons why Ashik's sample might be biased:

Problem	Solution
Ashik only selected a sample of 5 students to represent his entire school	Choose a **larger** sample. Ashik should have surveyed at least 20 or 30 students.
All the students in Ashik's sample were in one class, so were in the same year group.	Choose a **random sample**. If everyone has an equal chance of being selected there will be a range of ages.

You will need to use problem-solving skills throughout your exam – **be prepared!**

Worked example

Ashik wants to investigate the number of hours of television students in his school watch each week. He surveys five members of his class and gets these results:

15   6   22   11   18

(a) Use this data to estimate the mean number of hours of television watched each week by students in Ashik's school. **(2 marks)**

15 + 6 + 22 + 11 + 18 = 72
72 ÷ 5 = 14.4
An estimate for the mean of the population is 14.4 hours.

(b) Comment on the reliability of your estimate. **(1 mark)**

The estimate is not very reliable because of the small sample size.

(c) How could Ashik reduce bias in his sample? **(2 marks)**

Select a larger sample, and select a random sample from the whole population.

Now try this

There is more about probability on page 123.

Amy and Paul are investigating whether trains at their local station run on time. This table summarises their results.

	Amy	Paul
Number of trains observed	6	25
Number of trains late	3	8

Amy says 'There is a 50–50 chance that a train will be late.'

(a) Comment on the reliability of Amy's statement. **(1 mark)**

(b) Use Paul's results to estimate the probability of a train being late. **(1 mark)**

Stratified sampling

What is stratified sampling?

A stratified sample is one in which the population is split into groups. The number of members selected from each group for the sample is proportional to the size of that group.

There are twice as many boys as girls in this population...

Girls   Boys

... so you need twice as many boys as girls in a stratified sample.

Sampling fraction

Use this rule to find the sampling fraction for a stratified sample.

$$\text{Sampling fraction} = \frac{\text{Sample size}}{\text{Population size}}$$

You multiply the sampling fraction by the size of each group to work out how many members to select from that group.

In the example on the left the sampling fraction is $\frac{6}{24}$. So you need $8 \times \frac{6}{24} = 2$ girls and $16 \times \frac{6}{24} = 4$ boys in your stratified sample.

Worked example

The table below gives information about the members of a tennis club.

	Male	Female	Total
18 or over	74	66	140
Under 18	22	42	64
Total	96	108	204

Malik is carrying out a customer satisfaction survey. He chooses a sample of 30 members, stratified by age group and gender.

Work out the number of females under 18 he should include in his sample. **(3 marks)**

$42 \times \frac{30}{204} = 6.176...$

Malik should include 6 females under 18 in his sample.

There's a lot of information given so read the whole question carefully. Start by calculating the sampling fraction. The population size is the total in the bottom right of the table.

$$\text{Sampling fraction} = \frac{\text{Sample size}}{\text{Populataion size}}$$

$$= \frac{30}{204}$$

You need to use this in your calculation so you can leave it in this form. Remember that you can only select a **whole number** of students from each group, so you should round your answer to the nearest whole number.

Now try this

This table shows the number of employees at a large department store. The human resources manager wants to select a random sample of 40 employees, stratified by gender.

	Full-time	Part-time	Total
Male	80	36	116
Female	105	19	124
Total	185	55	240

(a) How many male employees should she select for her sample? **(3 marks)**

(b) Suggest **one** method she could use to select a random sample. **(1 mark)**

 Look at page 115 for methods of random sampling.

Capture-recapture

You can estimate the size of a large animal population using the **Peterson capture-recapture** method. This is sometimes called the **mark and recapture** method. The diagrams below show how you could use it to estimate the number of fish in a lake.

 Catch a sample of fish from the lake and mark them. Return the marked fish to the lake.

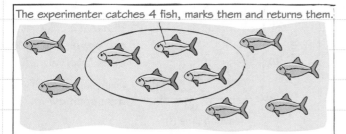

The experimenter catches 4 fish, marks them and returns them.

You can use equivalent fractions to estimate the size of the population. You **assume** that the following two fractions are equivalent.

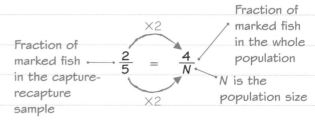

Fraction of marked fish in the capture-recapture sample

Fraction of marked fish in the whole population

$$\frac{2}{5} = \frac{4}{N}$$

×2

×2

N is the population size

 Later, catch a second sample of fish. Count how many of this sample are marked.

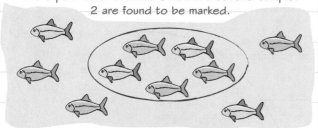

The experimenter catches 5 fish in a second sample. 2 are found to be marked.

Using a formula

You can use this formula to estimate the population size *N*.

$$N = \frac{Mn}{m}$$

LEARN IT!

M = number of fish marked then released

n = size of recapture sample

m = number of marked fish in recapture sample

Problem solved!

If you can't remember the formula for capture-recapture you can answer this question using equivalent fractions:

×3

$$\frac{4}{15} = \frac{12}{N}$$

×3

Always check that your answer **makes sense**. The estimate must be larger than the number of owls captured.

You will need to use problem-solving skills throughout your exam – **be prepared!**

Worked example

A conservation charity wants to estimate the number of owls in an area of woodland. A scientist captures 12 owls, marks them with a tag, and releases them. One month later she captures 15 owls. She finds that 4 of the owls are tagged. Estimate the total size of the owl population in this area. **(3 marks)**

$$N = \frac{Mn}{m}$$
$$= \frac{12 \times 15}{4} = 45$$

Now try this

Ravina wants to find an estimate for the number of eagles in a sanctuary. She catches a sample of 70 eagles in the sanctuary and tags each of them. These birds are then released back into the sanctuary.

The next day she catches a sample of 60 eagles in the sanctuary. 12 of these eagles are tagged.

Work out an estimate for the total number of eagles in the sanctuary. **(3 marks)**

Cumulative frequency

In your exam you might have to draw a cumulative frequency graph, or use one to find the median or the interquartile range.

How to draw a cumulative frequency graph

Reaction time t (s)	Frequency	Cumulative frequency
$0 < t \leq 0.1$	2	2
$0.1 < t \leq 0.2$	5	$2 + 5 = 7$
$0.2 < t \leq 0.3$	18	$7 + 18 = 25$
$0.3 < t \leq 0.4$	5	$25 + 5 = 30$
$0.4 < t \leq 0.5$	1	$30 + 1 = 31$

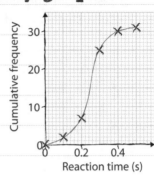

1. Plot 0 at the beginning of the first class interval.

2. Plot each value at the **upper** end of its class interval.

3. Join your points with a **smooth curve**.

Add a column for **cumulative frequency** to your frequency table.

Check that your final value is the same as the total frequency.

Here's another example:

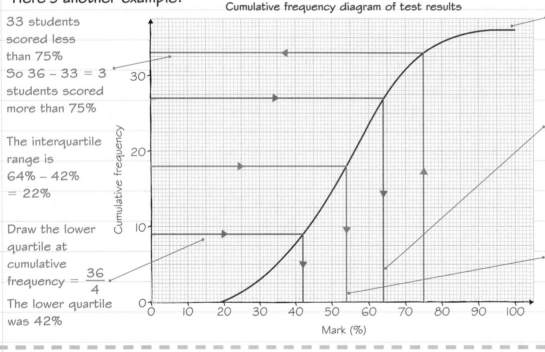

Cumulative frequency diagram of test results

33 students scored less than 75%
So $36 - 33 = 3$ students scored more than 75%

There were 36 students in the class. (This is the **first fact** you should establish.)

The interquartile range is
$64\% - 42\%$
$= 22\%$

Draw the upper quartile at cumulative frequency $= 3 \times \dfrac{36}{4}$
The upper quartile was 64%

Draw the lower quartile at cumulative frequency $= \dfrac{36}{4}$
The lower quartile was 42%

Draw the median at cumulative frequency $= \dfrac{36}{2}$
The median was 54%

Now try this

This cumulative frequency graph shows the reaction times recorded by a group of students in an experiment.

Estimate the median and interquartile range of the reaction times.   **(3 marks)**

Draw lines on your graph to show the values you are reading off.

Worked solution video

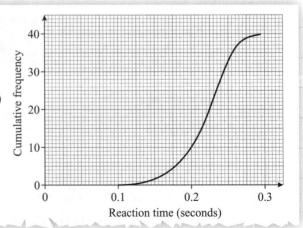

Reaction time (seconds)

118

Box plots

Box plots show the median, upper and lower quartiles, and the largest and smallest values of a set of data. They are often used to compare distributions.

Half the weights were between 60 kg and 78 kg.

25% of the weights were greater than 78 kg.

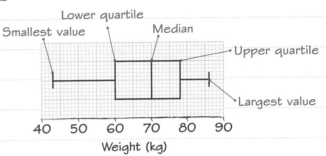

Smallest value   Lower quartile   Median   Upper quartile   Largest value

Weight (kg)

Drawing box plots

You might need to draw a box plot from data presented:

- as a list of values
- on a stem-and-leaf diagram
- in a cumulative frequency diagram.

To draw a box plot you **always** need to find the **least** and **greatest** values, the **quartiles** and the **median**.

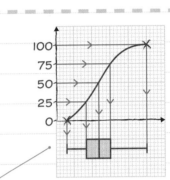

You can read the median and quartiles off a cumulative frequency diagram.

Worked example

The box plot gives information about the heights, in metres, of the trees in a park.

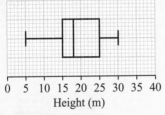

Height (m)

There are 162 trees in a park.
Estimate the number of trees which are less than 10 m tall. **(1 mark)**

12.5% of 175 = $\frac{12.5}{100} \times 162 = 20.25$

20 is a good estimate.

10 m is half-way between the smallest value and the lower quartile.

25% of the data values lie between the smallest value and the lower quartile.

So a good estimate is that 12.5% of the trees are less than 10 m tall.

Box plot checklist

Drawn on graph paper	✓
Ruler and sharp pencil	✓
Needs a scale	✓
Shows range and interquartile range	✓
Shows median and quartiles	✓

Now try this

Here is a table showing information about some test scores for 120 boys.

(a) Draw a box plot for this data. **(3 marks)**

(b) How many boys scored over 50? **(1 mark)**

(c) Estimate how many boys scored less than 20. **(1 mark)**

Lowest score	10
Lower quartile	29
Median	37
Interquartile range	21
Range	53

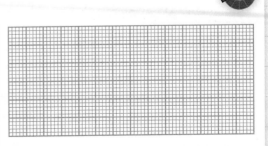

Histograms

Histograms are a good way to represent grouped data with **different** class widths.

Worked example

This table shows the finishing times in minutes of runners in a cross-country race.

Time (t minutes)	Frequency	Frequency density
$16 \leq t < 20$	12	3
$20 \leq t < 30$	45	4.5
$30 \leq t < 50$	28	1.4

Draw a histogram to represent the data.

(3 marks)

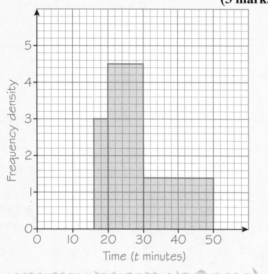

Histogram facts

No gaps between the bars. ✓

Area of each bar is proportional to frequency. ✓

Vertical axis is labelled 'Frequency density'. ✓

Bars can be different widths. ✓

Frequency density $= \dfrac{\text{frequency}}{\text{class width}}$ ✓

The class widths in the frequency table are **different widths** so a histogram is the most suitable graph to draw.

1. Calculate the frequency densities.

   $\dfrac{12}{4} = 3$ $\dfrac{45}{10} = 4.5$ $\dfrac{28}{20} = 1.4$

2. Add these values to the table as an extra column.

3. Label the vertical axis 'Frequency density'.

4. Choose a scale for the vertical axis and draw each bar.

Area and estimation

You can use the area under a histogram to estimate frequencies. An estimate for the **number** of maggots between 1 mm and 2 mm long is:

$0.5 \times 22 + 0.5 \times 6 = 14$

You might need to answer proportion questions about histograms in your exam. The total frequency is:

$1 \times 14 + 0.5 \times 22 + 1.5 \times 6 = 34$

So an estimate for the **proportion** of maggots between 1 mm and 2 mm long is: $\dfrac{14}{34} = 0.4117...$ or 41% (2 s.f.)

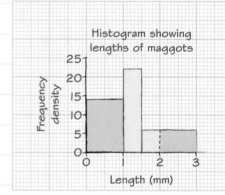

Now try this

A speed camera recorded the speed of some vehicles on a motorway. The table on the right shows the results.

(a) On graph paper, draw a histogram to illustrate this data. **(3 marks)**

(b) Estimate the proportion of vehicles travelling at more than 65 mph. **(2 marks)**

Speed, s (mph)	Frequency
$0 < s \leq 40$	48
$40 < s \leq 50$	32
$50 < s \leq 70$	128
$70 < s \leq 80$	88
$80 < s \leq 110$	24

Frequency polygons

You can represent grouped data using a **frequency polygon**. Look at this example.

Reaction time (r milliseconds)	Frequency
$100 \leq r < 200$	7
$200 \leq r < 300$	15
$300 \leq r < 400$	10

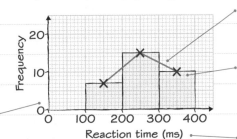

Join the points with **straight lines**. Make sure you use a ruler.

Plot points at the **midpoint** of each class interval.

You always record **frequency** on the vertical axis.

If you draw a histogram on the same graph the frequency polygon joins together the midpoints of the tops of the bars.

This frequency polygon shows the reaction times of a class of students.

Worked example

30 students timed how long it took them to complete a jigsaw puzzle. The results were recorded in a grouped frequency table:

Time (t minutes)	Frequency	Midpoint
$10 \leq t < 14$	2	12
$14 \leq t < 18$	5	16
$18 \leq t < 22$	12	20
$22 \leq t < 26$	8	24
$26 \leq t < 30$	3	28

Show this information on a frequency polygon.   **(3 marks)**

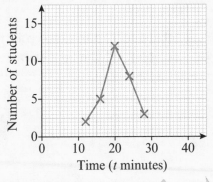

Start by working out the midpoints of the class intervals.

The midpoint of the class interval $10 \leq t < 14$ is $\frac{10 + 14}{2} = 12$

Check it!

In your exam you will only be asked to draw a frequency polygon for data with **equal class intervals**. So make sure that your midpoints are the same distance apart.

Estimating the mean

You can estimate the mean for data given in a frequency polygon using this formula:

$$\text{Mean} \approx \frac{\text{Sum of (midpoint} \times \text{frequency)}}{\text{Total frequency}}$$

For the worked example on the left:

$$\frac{(2 \times 12) + (5 \times 16) + (12 \times 20) + (8 \times 24) + (3 \times 28)}{30} = 20.6666...$$

An estimate for the mean time is 20 minutes, 40 seconds.

There is more about estimating the mean of grouped data on page 111.

Now try this

This table shows the times taken, in minutes, for 50 people to solve a crossword puzzle.

(a) Draw a frequency polygon for this data.
    (3 marks)

(b) Which interval contains the median time?
    (1 mark)

Time, t (minutes)	Frequency
$0 < t \leq 10$	3
$10 < t \leq 20$	9
$20 < t \leq 30$	11
$30 < t \leq 40$	18
$40 < t \leq 50$	7
$50 < t \leq 60$	2

Comparing data

You can use averages like the **mean** or **median** and measures of spread like the **range** and **interquartile range** to compare two sets of data. Follow these steps.

1 Calculate an average and a measure of spread for both data sets.

2 Write a sentence for each statistic, **comparing** the values for each data set.

3 Only make a statement if you can back it up with **statistical evidence**.

Calculate then compare

You will often have to compare two sets of data presented in different ways. Make sure you calculate the **same** statistics for both data sets. You will get marks for calculating the statistics correctly **and** for comparing the data sets.

Worked example

These box plots show information about the prices of used cars at two different garages.

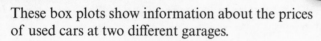

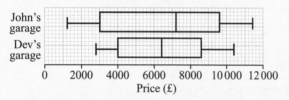

John's garage
Dev's garage

Price (£)

Compare the prices of used cars at the two garages.   **(2 marks)**

	Median (£)	IQR (£)
John's	7200	9600 − 3000 = 6600
Dev's	6400	8600 − 4000 = 4600

The cars at Dev's garage were cheaper on average (smaller median).

The prices of the cars at John's garage were more spread out (larger IQR).

Remember you need to mention one average **and** one measure of spread. You should use the **interquartile range** as your measure of spread if the data is presented on a box plot or cumulative frequency diagram.

Start by writing the median and IQR for each set of data. Then write a sentence comparing each statistic.

Golden rules

1 When you are comparing two sets of data you should use **one average**, like the median, and **one measure of spread**, like the interquartile range.

2 **Interpret** your results in the **context** of the question.

Now try this

(a) The cumulative frequency graph shows some information about the times taken for 40 girls to complete a challenge. The shortest time taken was 8 minutes.
Draw a box plot for this data.   **(3 marks)**

(b) The box plot shows some information about the times taken by 40 boys to complete the same challenge.

Make **two** comments to compare the times taken by the girls and the times taken by the boys to complete the challenge.   **(2 marks)**

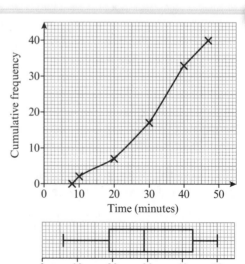

Cumulative frequency

Time (minutes)

Time (minutes)

Probability

For **equally likely outcomes** the probability (P) that something will happen is:

Probability = $\dfrac{\text{number of successful outcomes}}{\text{total number of possible outcomes}}$

If you know the probability that an event **will** happen, you can calculate the probability that it won't happen:

P(Event doesn't happen) = 1 − P(Event happens)

 The probability of rolling a 6 on a normal fair dice is $\frac{1}{6}$. So the probability of **not** rolling a 6 is $1 - \frac{1}{6} = \frac{5}{6}$

Add or multiply?

Events are **mutually exclusive** if they can't **both** happen at the same time. For mutually exclusive events:

P(A or B) = P(A) + P(B)

Events are **independent** if the outcome of one doesn't affect the outcome of the other. For independent events:

P(A and B) = P(A) × P(B)

Worked example

Amir designs a game for his school fete. This table shows the probability of winning a prize.

Prize	Badge	Keyring	Cuddly toy
Probability	0.35	0.18	0.07

(a) What is the probability of **not** winning a prize? **(2 marks)**

P(Win) = 0.35 + 0.18 + 0.07 = 0.6
P(Not win) = 1 − 0.6 = 0.4

(b) Amir plays the game three times. What is the probability that he does not win a prize? **(2 marks)**

0.4 × 0.4 × 0.4 = 0.064

(a) To work out the probability of winning **any** prize you need to add together the probabilities. Then you can use this rule to work out the probability of not winning a prize:

$P\left(\begin{array}{c}\text{Not winning}\\\text{a prize}\end{array}\right) = 1 - P\left(\begin{array}{c}\text{Winning a}\\\text{prize}\end{array}\right)$

(b) To work out the probability of Amir not winning on any of his three games, you need to multiply the probabilities.

Sample space diagrams

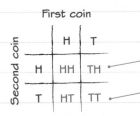

A **sample space diagram** shows you all the possible outcomes of an event. Here are all the possible outcomes when two coins are flipped.

There are four possible outcomes. TH means getting a tail on the first coin and a head on the second coin.

The probability of getting two tails when two coins are flipped is $\frac{1}{4}$ or 0.25. There are 4 possible outcomes and only 1 successful outcome (TT).

Now try this

The table shows the probabilities of different delivery times for a first class letter.

Delivery time	Next day	1 day late	More than 1 day late
Probability	45x	3x	2x

Work out the probability of a first class letter arriving one day late. **(3 marks)**

 Don't just find x. You need to answer the question by writing the probability of the letter arriving one day late.

 Worked solution video

Relative frequency

You need to be able to calculate probabilities for data given in graphs and tables. You can use this formula to estimate a probability from a frequency table:

$$\text{Probability} = \frac{\text{frequency of outcome}}{\text{total frequency}}$$

When a probability is calculated like this it is sometimes called a **relative frequency**.

Golden rule

Probability estimates based on relative frequency are **more accurate** for larger samples (or for a larger number of trials in an experiment).

In the sample there were 15 + 10 = 25 eggs which weighed 55 g or more. So an estimate for the probability of picking an egg that weighs 55 g or more is $\frac{25}{40}$ or $\frac{5}{8}$.

Worked example

An egg farm weighed a sample of 40 eggs. It recorded the results in a frequency table.

Weight, w (g)	Frequency
$45 \leqslant w < 50$	6
$50 \leqslant w < 55$	9
$55 \leqslant w < 60$	15
$60 \leqslant w < 65$	10

(a) Roselle buys some eggs from the farm and picks one at random. Estimate the probability that the egg weighs 55 g or more. **(2 marks)**

$P(w \geqslant 55) \approx \frac{25}{40} = \frac{5}{8}$

(b) Comment on the accuracy of your estimate. **(1 mark)**

40 is a fairly small sample size, so the estimate is not very accurate.

Experimental probability

You can carry out an experiment to estimate the probability of something happening. This table shows the results of throwing a drawing pin 60 times.

Number of trials	10	20	30	40	50	60
Frequency of landing point up	8	11	17	25	30	37

To estimate the probability that the drawing pin will land point up, you calculate the relative frequency. The most accurate estimate will be based on the largest number of trials.

Now try this

A four-sided dice is rolled 40 times. The results are shown in the table.

Number	1	2	3	4
Frequency	8	4	21	7

Worked solution video

(a) Work out the estimated probability of getting a 3. **(1 mark)**

(b) Work out the theoretical probability of getting a 3 on a fair four-sided dice. **(1 mark)**

(c) Do you think this dice is fair? Give a reason for your answer. **(1 mark)**

If the dice was fair, then the experimental probability would get closer to the theoretical probability as the number of trials increased.

Venn diagrams

You can use a Venn diagram to show frequencies in a probability question. This Venn diagram shows the results when 50 people were asked whether they owned a dog (D) or a cat (C). The rectangle represents everyone who was surveyed. The **number** in each section tells you **how many** people that section represents.

In total 21 + 8 + 6 + 15 = 50 people were surveyed. This symbol represents all of them.

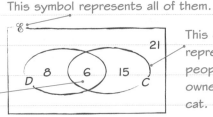

This oval represents people who owned a cat.

Some people owned a cat *and* a dog so the ovals overlap.

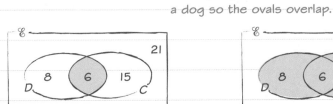

6 people owned a dog **and** a cat. You can write this as $D \cap C$. $\cap$ means **and** or **intersection**.

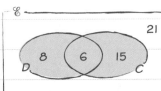

8 + 6 + 15 = 29 people owned a dog **or** a cat. You can write this as $D \cup C$. $\cup$ means **or** or **union**.

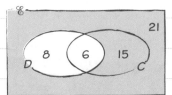

15 + 21 = 36 people **did not** own a dog. You can write this as D'. D' means **not** D or the **complement** of D.

Worked example

36 members of a youth club were surveyed about the sports they played.

19 members played tennis.

14 members played football.

6 members played both tennis and football.

(a) Draw a Venn diagram to show this information. **(3 marks)**

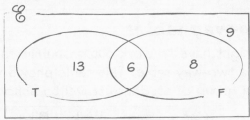

(b) One of the members is chosen at random. Write down the probability that the member plays neither tennis nor football. **(1 mark)**

$$\frac{9}{36} = \frac{1}{4}$$

$T \cap F$
Fill in the centre of the Venn diagram first. 6 members play both sports.

$T \cap F'$
The 19 members who play tennis include the 6 members who play both sports. So 19 − 6 = 13 members play tennis but *not* football.

$T' \cap F$
14 − 6 = 8 members play football but *not* tennis.

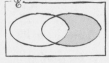

$T' \cap F'$
36 − 13 − 6 − 8 = 9 members play neither sport.

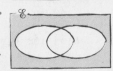

Check it!
13 + 6 = 19 members play tennis ✓
8 + 6 = 14 members play football ✓
13 + 6 + 8 + 9 = 36 members surveyed ✓

Now try this

The Venn diagram shows information about how 30 people in an office travel to work.

One person is chosen at random. Find the probability that this person

(a) uses both the bus and the train on their journey **(1 mark)**

(b) uses the bus at some point in their journey. **(2 marks)**

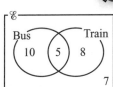

Conditional probability

If one event **has already occurred**, the probability of other events occurring might **change**. This is called conditional probability. The probability that an event X occurs **given that** an event Y has **already** occurred is written as P(X|Y).

Using Venn diagrams

You can solve some conditional probability problems using a Venn diagram. If an event has already occurred, then the sample space for the other events is **restricted**. These Venn diagrams show the outcomes of two events, A and B.

Complete sample space

$$P(A) = \frac{8 + 3}{8 + 3 + 4 + 5}$$

$$= \frac{11}{20}$$

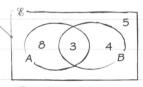

Event B occurs

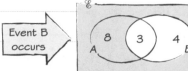

Restricted sample space **given that** event B has occurred

$$P(A \mid B) = \frac{3}{3 + 4} = \frac{3}{7}$$

This Venn diagram shows the burger toppings chosen by a group of 50 diners at a restaurant. The choices are avocado (A), bacon (B) and cheese (C).

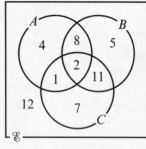

A diner is chosen at random.

(a) Given that the diner chooses bacon, find the probability that she also chooses avocado.   **(2 marks)**

$$\frac{8 + 2}{8 + 2 + 11 + 5} = \frac{10}{26} = \frac{5}{13}$$

A second diner is chosen at random.

(b) Given that the diner chooses at least one of the three toppings, find the probability that she chooses all three.   **(3 marks)**

$$\frac{2}{4 + 8 + 5 + 1 + 2 + 11 + 7} = \frac{2}{38} = \frac{1}{19}$$

(a) Look at this restricted sample space:

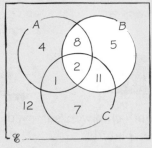

Of the 8 + 2 + 11 + 5 = 26 diners who chose bacon, 8 + 2 = 10 also chose avocado.

You will need to use problem-solving skills throughout your exam – **be prepared!**

Two-way tables

You might have to work out probabilities from a two-way table. This table shows how a group of students travel to school.

	Car	Bus	Walk	Bike
Male	4	17	20	9
Female	8	20	29	11

Given that a student bikes to school, the probability that they are male is $\frac{9}{9 + 11} = \frac{9}{20}$

---

1 Look at the Venn diagram in the worked example above. A third diner is chosen at random. Given that the diner does **not** choose bacon, find the probability that she chooses cheese.   **(3 marks)**

2 Look at the two-way table in the blue box above. A student is chosen at random from this group. Given that the student is male, find the probability that he walks to school.   **(3 marks)**

Tree diagrams

You can use a tree diagram to answer questions involving **conditional probability**.

A tree diagram shows all the possible outcomes from a series of events and their probabilities.

This is a tree diagram for Holly's journey to school.

You write the outcomes at the ends of the branches.
You can use shorthand like this.

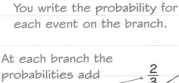 You write the probability for each event on the branch.

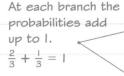 At each branch the probabilities add up to 1.
$\frac{2}{3} + \frac{1}{3} = 1$

The outcome of the first event can affect the probability of the second. Holly is less likely to be on time if she misses the bus.

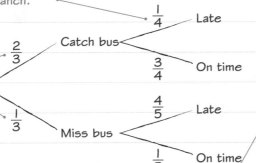

Outcome	Probability
CL	$\frac{2}{3} \times \frac{1}{4} = \frac{2}{12} = \frac{1}{6}$
CO	$\frac{2}{3} \times \frac{3}{4} = \frac{6}{12} = \frac{1}{2}$
ML	$\frac{1}{3} \times \frac{4}{5} = \frac{4}{15}$
MO	$\frac{1}{3} \times \frac{1}{5} = \frac{1}{15}$

Each branch is like a different parallel universe. In this universe, Holly catches the bus and gets to school on time.

You multiply along the branches to find the probability of each outcome.
The probability that Holly misses the bus and is late for school is $\frac{4}{15}$

Golden rules

1 Look out for the words **replace** or **put back** in a probability question.

With replacement: probabilities stay the same.

Without replacement: first probability stays the same while the others change.

2 MULTIPLY ALONG THE BRANCHES   ADD UP THE OUTCOMES

 Worked example

There are 3 strawberry yoghurts and 4 pineapple yoghurts in a fridge. Noah picks two yoghurts at random. Work out the probability that both the yoghurts are the same flavour.   **(4 marks)**

This is an example of selection **without replacement**. The two events are not independent. The probabilities for the second pick change depending on which flavour yoghurt was picked first. A tree diagram is the **safest** way to answer questions like this.

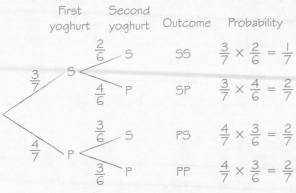

First yoghurt	Second yoghurt	Outcome	Probability
$\frac{3}{7}$ S	$\frac{2}{6}$ S	SS	$\frac{3}{7} \times \frac{2}{6} = \frac{1}{7}$
	$\frac{4}{6}$ P	SP	$\frac{3}{7} \times \frac{4}{6} = \frac{2}{7}$
$\frac{4}{7}$ P	$\frac{3}{6}$ S	PS	$\frac{4}{7} \times \frac{3}{6} = \frac{2}{7}$
	$\frac{3}{6}$ P	PP	$\frac{4}{7} \times \frac{3}{6} = \frac{2}{7}$

P (both yoghurts same flavour) = P(SS) + P(PP)
$= \frac{1}{7} + \frac{2}{7} = \frac{3}{7}$

 **Now try this**

Worked solution video

The probability that a student at Jen's school has a dog is 0.3.
If a student has a dog, the probability that they have a cat is 0.12.
If a student does not have a dog, the probability that they have a cat is 0.25.
A student is chosen at random.
Work out the probability that they do not have a cat.   **(4 marks)**

Problem-solving practice 1

Throughout your Higher GCSE exam you will need to **problem-solve**, **reason**, **interpret** and **communicate** mathematically. If you come across a tricky or unfamiliar question in your exam you can try some of these strategies:

- ☑ Sketch a diagram to see what is going on.
- ☑ Try the problem with smaller or easier numbers.
- ☑ Plan your strategy before you start.
- ☑ Write down any formulae you might be able to use.
- ☑ Use x or n to represent an unknown value.

---

1 An internet service provider advertises an internet connection speed of 12 Mbps. Dhevan records the speed of 8 internet connections at different distances from the phone exchange.

Distance (km)	9.2	13.6	7.5	4.2	5.9	5.3	22.1	0.8
Speed (Mbps)	8.3	7.8	8.5	11.4	10.3	9.5	6.2	10.9

(a) Use this information to predict the connection speed at a distance of 30 km from the exchange. **(4 marks)**

(b) Make two comments explaining why your prediction in part (a) may not be reliable. **(2 marks)**

Scatter graphs page 114

You will need graph paper to answer this question. You will get marks for choosing an appropriate graph and axes, and for showing your method clearly and neatly. For part (b), make sure you give two **different** reasons.

TOP TIP

Make sure you draw any lines of best fit with a ruler, and draw lines to show which values you are reading off your graph.

---

2 Huan spins this spinner. He keeps spinning until he lands on the unshaded sector.

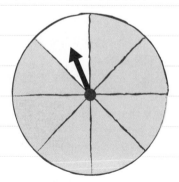

Work out the probability that Huan spins the spinner

(a) exactly twice **(2 marks)**

(b) more than twice. **(2 marks)**

Probability page 123

You might need to think carefully before working out which probabilities to multiply.

(a) Huan stops when he lands on white, so in order to spin the spinner **exactly** twice he must land on blue on the first spin, **and** white on the second spin.

(b) In order for Huan to spin more than twice he must land on blue on **both** of the first two spins.

TOP TIP

Use notation like P(Blue) and P(White) in probability questions to show your working clearly.

Problem-solving practice 2

3 Amy picks one card at random from group A and one card at random from group B.

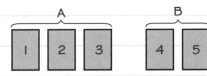

She adds together the numbers on the two cards. What is the probability that the total of the two cards is

(a) 8   (b) 7   **(3 marks)**

Probability page 123
Counting strategies page 13

You can tackle this in two different ways.

1. Consider all the possible outcomes of picking two cards. Write a list or use a sample space diagram.
2. Consider the two picks as two separate independent events, and multiply the probabilities of each.

TOP TIP

In a probability question make sure you have considered all the possible outcomes, and all the successful outcomes.

4 This table and histogram show the batting averages of the members of England's 2009 Ashes cricket squad. The vertical axis has not been numbered.

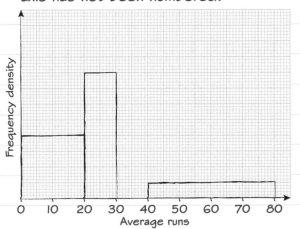

Average runs, x	Frequency
$0 \leqslant x < 20$	4
$20 \leqslant x < 30$	
$30 \leqslant x < 40$	5
$40 \leqslant x < 80$	

Complete the table and the histogram. **(4 marks)**

Histograms page 120

You can use area to answer this question. In a histogram, area is directly proportional to frequency. You are given the bar for the $0 \leqslant x < 20$ class interval. Use this to work out what frequency each large square on the graph represents. You can use this information to fill in the missing frequencies and draw the missing bar.

TOP TIP

You can use this rule in histogram questions:

$$\text{Frequency} = \frac{\text{frequency}}{\text{density}} \times \frac{\text{class}}{\text{width}}$$

5 In a board game, Diego picks a question card. He can pick an easy (E), medium (M) or hard (H) question.

The probabilities of picking each type of question are 0.54, 0.31 and 0.15 respectively.

The probabilities that he gets each type of question correct are 0.8, 0.5 and 0.1 respectively.

Work out the probability that Diego gets his question correct.   **(5 marks)**

Tree diagrams page 127

Either draw a tree diagram, or write down the successful outcomes and work out their probabilities. For example:

P(E, correct) = 0.54 × 0.8 = 0.432

TOP TIP

If you draw a tree diagram for a conditional probability question you will probably get some or all of the marks for this question because you won't leave out any of the possibilities.

Answers

NUMBER

1. Factors and primes

1. (a) $2^2 \times 5 \times 7^2$ (b) 28
2. (a) $3^2 \times 7 = 63$ (b) $2 \times 3^5 \times 5 \times 7^2 = 119\,070$

2. Indices 1

1. (a) 7^8 (b) $n = 3$
2. $k = 2$
3. (a) (i) 2^4 (ii) 7^{12} (iii) 5^2
(b) $n = 11$

3. Indices 2

1. (a) 7^{h-k} (b) 7^{2h} (c) 7^{h+2k}
2. $n = -3$
3. $\sqrt{\dfrac{49}{7^3}} = \sqrt{\dfrac{7^2}{7^3}} = \sqrt{7^{-1}} = 7^{-\frac{1}{2}}$

4. Calculator skills 1

(a) 4.197 530 864 (b) 4.20 (3 s.f.)

5. Fractions

1. (a) $\frac{9}{20}$ (b) $5\frac{5}{18}$ (c) $1\frac{4}{5}$ (d) 5
2. (a) $\frac{9}{20}$ (b) 120 cl

6. Decimals

1. $\frac{9}{20}, \frac{1}{2}, 0.55, 0.6$
2. e.g. $\dfrac{1}{250} = \dfrac{1}{2 \times 5^3}$. Only factors in denominator are 2 and 5 so terminating decimal.
3. e.g. $\dfrac{1}{140} = \dfrac{1}{2^2 \times 5 \times 7}$. Factors in denominator other than 2 and 5 so recurring decimal.
4. $0.454\,545\ldots = 0.\dot{4}\dot{5}$

7. Estimation

1. $\dfrac{80 \times 300}{60 \times 40} = 10$
2. (a) 108 (or 81, rounding $\frac{4}{3}$ to 1)
(b) $3 < 3.14$ and $3 < 3.2$ (and $1 < \frac{4}{3}$) so underestimate

8. Standard form

1.8×10^{-9} g

9. Recurring decimals

1. Let $n = 0.545\,454\,54\ldots$
$100n = 54.545\,454\ldots$
$99n = 54$
$n = \frac{54}{99} = \frac{6}{11}$
2. Let $n = 0.018\,181\,18\ldots$
$100n = 1.818\,181\,818\ldots$
$99n = 1.8$
$n = \frac{1.8}{99} = \frac{1}{55}$
3. Let $n = 0.351\,351\,351\ldots$
$1000n = 351.351\,351\,351\ldots$
$999n = 351$
$n = \frac{351}{999} = \frac{13}{37}$

10. Upper and lower bounds

1. LB for width = LB for area ÷ UB for length
$= 315 \div 22.5$
$= 14$ cm
2. 13.5 m^2

11. Accuracy and error

1.

	Upper bound	Lower bound
Distance	425.5 m	424.5 m
Time	85.5 seconds	84.5 seconds

UB for speed $= \frac{425.5}{84.5} = 5.035\,50\ldots$
LB for speed $= \frac{424.5}{85.5} = 4.964\,91\ldots$
Speed = 5 m/s (1 s.f.) because UB and LB both round to 5 m/s to 1 s.f.

2.

	270 m	310 m	500 m²
Lower bound	265 m	305 m	450 m²
Upper bound	275 m	315 m	550 m²

UB for Area $= \frac{1}{2} \times 275 \times 315 \times \sin 80°$
$= 42\,654.4858\ldots$
$42\,654.4858\ldots \div 450 = 94.787\,74\ldots$
The farmer might need 95 bags so he does not definitely have enough.

12. Surds 1

1. $\sqrt{32} + \sqrt{98} = \sqrt{16 \times 2} + \sqrt{49 \times 2}$
$= 4\sqrt{2} + 7\sqrt{2}$
$= 11\sqrt{2}$
$p = 11$
2. $\dfrac{35}{\sqrt{7}} = \dfrac{35\sqrt{7}}{7} = 5\sqrt{7}$
3. $x = 32$

13. Counting strategies

3.78×10^{10}

14. Problem-solving practice 1

1. 96 mm
2. $2 \times \frac{3}{8} = \frac{3}{4}$ of a bag per day.
$14 \div \frac{3}{4} = 18\frac{2}{3}$. Susan can feed the dogs for 18 days from 1 bag.
3. $n = 0.928\,282\,8\ldots$
$100n = 92.828\,282\,8\ldots$
$99n = 91.9$
$n = \frac{91.9}{99} = \frac{919}{990}$

15. Problem-solving practice 2

4. (a) 2×10^n (b) $1.6 \times 10^{4m+1}$
5. 178 cm (3 s.f.)
6. (a) 625 (b) $\frac{1310}{17\,576}$ or 0.0745 (3 s.f.)

ALGEBRA

16. Algebraic expressions

1. h^{12}
2. (a) $16a^{20}b^4$ (b) $15x^7y^9$ (c) $3d^6g^5$
3. (a) $4p^5$ (b) $\frac{1}{4x^3y^{\frac{2}{3}}}$

17. Expanding brackets

1. (a) $x^2 + 4x - 5$
   (b) $p^2 - 12p + 36$
2. (a) $x^3 + 12x^2 + 27x$
   (b) $n^3 + 11n^2 + 39n + 45$

18. Factorising

1. (a) $2(2a - 3)$ (b) $y(y + 5)$
2. (a) $3g(4 + g)$ (b) $(p - 14)(p - 1)$ (c) $2x(3x - 4y)$
3. (a) $4ma(1 - 6m)$ (b) $(p + 8)(p - 8)$
4. $(3x - 2)(x - 2)$

19. Linear equations 1

1. (a) $w = 7$ (b) $x = -1$
2. (a) $y = -\frac{7}{4}$ (b) $m = 4$

20. Linear equations 2

1. (a) $w = -5$ (b) $x = 3$
2. (a) $y = 6$ (b) $m = 37$

21. Formulae

21.9 cm

22. Arithmetic sequences

(a) nth term $= 4n - 1$
(b) No. 22nd term $= 87$, 23rd term $= 91$
   (or $4n - 1 = 89$
           $n = 22.5$ which is not a whole number)

23. Solving sequence problems

1. (a) 49, 46, 41   (b) -14 (8th term)
2. $a = -10$, $b = -2$
3. (a) $52 = 3n - 1$
       $n = 17\frac{2}{3}$ so 52 is not a term in the sequence
   (b) $u_n = u_{n-1} + 3$
   (c) 26 (9th term)

24. Quadratic sequences

$2n^2 - 2n + 5$

25. Straight-line graphs 1

$y = -2x + 2$

26. Straight-line graphs 2

1. $y = 6x - 20$
2. $y = 2x - 1$
3. $k = 6$

27. Parallel and perpendicular

1. $y = \frac{1}{2}x + 4$
2. (a) $y = \frac{4}{3}x$
   (b) $y = -\frac{3}{4}x + 12.5$

28. Quadratic graphs

(a)

x	-3	-2	-1	0	1	2	3
y	11	6	3	2	3	6	11

(b)

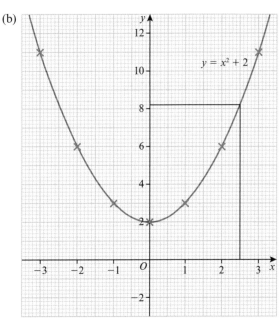

$y = x^2 + 2$

(c) $y = 8.2$ (to the nearest small square)

29. Cubic and reciprocal graphs

1. (a) D   (b) E   (c) A   (d) B   (e) F   (f) C

30. Real-life graphs

(a) 30 minutes
(b) 30 km/h

31. Quadratic equations

(a) $m = 6$, $m = 2$ (b) $w = -4$, $w = 9$
(c) $y = \frac{3}{5}$, $y = -8$ (d) $x = \frac{5}{7}$, $x = -1$

32. The quadratic formula

1. (a) $x = -1.16$, $x = 0.74$ (2 d.p.)
   (b) $m = -106.39$, $m = 56.39$ (2 d.p.)
2. $x = 6.42$, $x = 0.0779$ (3 s.f.)

33. Completing the square

1. (a) $x = -5 \pm \sqrt{13}$
   (b) $y = 3 \pm 2\sqrt{6}$ (or $y = 3 \pm \sqrt{24}$)
2. $2(x + 5)^2 - 43$

34. Simultaneous equations 1

1. $x = 5$, $y = 1.5$
2.

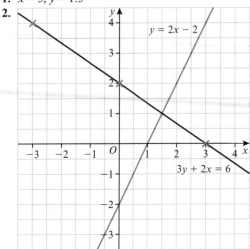

$y = 2x - 2$

$3y + 2x = 6$

$x = 1.5$, $y = 1$

35. Simultaneous equations 2

1. $x = 5$, $y = 1$ and $x = -3$, $y = -3$

2. $(2, 3)$ and $\left(-\frac{6}{5}, -\frac{17}{5}\right)$

36. Equation of a circle

(a), (b)

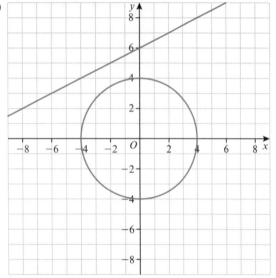

(c) The graphs of $x^2 + y^2 = 16$ and $2y - x = 12$ do not intersect, so the simultaneous equations have no solutions.

37. Inequalities

1. (a) $n > -2$ (b) $n \geqslant 22$

2. $x = 8$

3. $n = -2, -1, 0, 1$

4. $x < \frac{1}{8}$

38. Quadratic inequalities

1. $-2 < p < 2$

2. $-2 < x < 7$

39. Trigonometric graphs

(a)

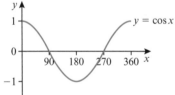

(b) $315°$

40. Transforming graphs

(a) $(5, -4)$ (b) $(2, -9)$ (c) $(2, 4)$ (d) $(-2, -4)$

41. Inequalities on graphs

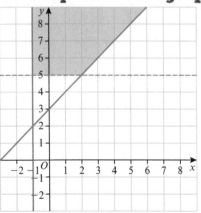

42. Using quadratic graphs

(a)

x	-5	-4	-3	-2	-1	0	1	2
y	7	1	-3	-5	-5	-3	1	7

(b)

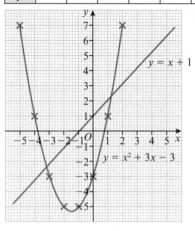

43. Turning points

(a) $(0, -3)$ (b) $(2, -7)$ (c) $x = -3.8$, $x = 0.8$

(d) $x = -3.2$, $x = 1.2$

44. Sketching graphs

(a)

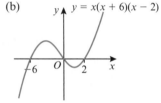

(b)

$y = x(x + 6)(x - 2)$

45. Iteration

(a) $x + \sqrt{x} - 5 = 0$ $(+ 5)$

    $x + \sqrt{x} = 5$ $(-\sqrt{x})$

         $x = 5 - \sqrt{x}$

(b) 3.209 (3 d.p.)

46. Rearranging formulae

1. $t = \dfrac{4p + 1}{3}$

2. $w = \dfrac{m^2 - 7}{5}$

3. $P = \dfrac{100}{Q^2 + 5}$

47. Algebraic fractions

1. $\dfrac{2a + 9}{6}$

2. $x = \dfrac{1}{120}$

3. (a) $\dfrac{x(x - 2)}{x + 5}$ (b) $\dfrac{m - 6}{3m^2}$

48. Quadratics and fractions

1. $x = -\frac{5}{3}$, $x = 2$

2. $x = \frac{3}{4}$, $x = -3$

3. $x = -\frac{15}{2}$, $x = -4$

49. Surds 2

1. $(3 - \sqrt{12})^2 = (3 - \sqrt{12})(3 - \sqrt{12})$

                $= 9 - 3\sqrt{12} - 3\sqrt{12} + (\sqrt{12})^2$

                $= 21 - 6\sqrt{12}$

                $= 21 - 6 \times 2\sqrt{3}$

                $= 21 - 12\sqrt{3}$

2. $x = 3$ and $k = 8$

50. Functions
1. (a) $f(1) = 4$ (b) $b = 1.5$
2. (a) $gf(x) = 4x^2 - 8x$ (b) $x = 0, x = 2$

51. Inverse functions
1. (a) $f^{-1}(x) = \dfrac{x + 1}{2}$ (b) 7
2. $g^{-1}(x) = \dfrac{6}{x - 1}$

52. Algebraic proof
1. e.g. $\frac{1}{2}n(n + 1) + \frac{1}{2}(n + 1)((n + 1) + 1) = \frac{1}{2}n(n+1) + \frac{1}{2}(n+1)(n+2)$
$= \frac{1}{2}(n +1)(n + (n + 2))$
$= \frac{1}{2}(n +1)(2n + 2)$
$= \frac{1}{2}(n + 1) \times 2(n + 1)$
$= (n + 1)(n + 1)$
$= (n + 1)^2$

2. e.g. $(n + 4)^2 - n^2 = n^2 + 8n + 16 - n^2$
$= 8n + 16$
$= 8(n + 2)$

The mean of $(n + 4)$ and $n = \dfrac{n + 4 + n}{2} = n + 2$

Therefore, the difference between the squares, $8(n + 2)$, is eight times the mean, $(n + 2)$

53. Exponential graphs
(a) 80   (b) 51.2 (or 51)
(c)

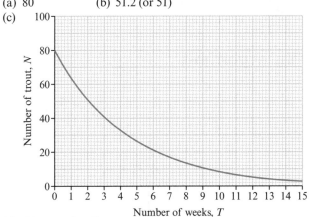

(d) 6 (accept 6 or 7)

54. Gradients of curves
(a) 0.8 cm/s (accept 0.7–0.9)   (b) 0.5 cm/s (accept 0.4–0.6)

55. Velocity–time graphs
(a)

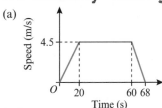

(b) 243 m

56. Areas under curves
34

57. Problem-solving practice 1
1. nth term = $6n + 3$
$6 \times 10 + 3 = 63$
$6 \times 11 + 3 = 69$
So 65 is not a term in the sequence.
2. 3 minutes $= \frac{3}{60} = \frac{1}{20}$ hours
Speed $= \dfrac{\text{distance}}{\text{time}} = y \div \frac{1}{20} = y \times \frac{20}{1} = 20y$

3. (a) $y = 8x - 30$
(b) When $x = 10$: $y = 8 \times 10 - 30 = 50$, so (10, 50) is a point on the line.

58. Problem-solving practice 2
4. 750
5. $a = 9$
6. L intersects C when these simultaneous equations are satisfied:
$x^2 + y^2 = 20$ (1)
$y = 2x - 10$ (2)
Substitute (2) into (1)
$x^2 + (2x - 10)^2 = 20$
$5x^2 - 40x + 80 = 0$
$x^2 - 8x + 16 = 0$
$(x - 4)^2 = 0$
$x = 4$
Only one solution, so only one point of intersection. Hence L is a tangent to C.

RATIO & PROPORTION
59. Calculator skills 2
1. 86.3%
2. Aisha: £557.50; Joshua: £540; Aisha spends the most on rent.

60. Ratio
1. £161
2. (a) 21
   (b) 24

61. Proportion
e.g. €16.55 = £14.39
1.5 lb = 0.681 81… kg
Cost in £ per kg in France = 14.39 ÷ 1.25 = 11.513
Cost in £ per kg in England = 8.97 ÷ 0.681… = 13.156
The cheese is cheaper in France

62. Percentage change
1. 34.6% (1 d.p.)
2. 59.6% (1 d.p.)

63. Reverse percentages
1. £45
2. £220 000

64. Growth and decay
1. 230 300 (nearest 100)
2. (a) £5304.50
   (b) 4 years

65. Speed
86.4 km/h

66. Density
9.234 g/cm³

67. Other compound measures
1. 16.6 km/l
2. 2.4 minutes or 2 minutes and 24 seconds

68. Proportion and graphs

(a) 4 inches   (b) 32 cm
(c) Graph is a straight line and passes through the origin (0, 0)

69. Proportionality formulae

1. (a) $x = 7.2y$ (b) 230.4   (c) 1.25
2. 8 cm

70. Harder relationships

1. (a) $t = 0.92\sqrt{d}$
    (b) 5.98 seconds
2. 3.09 km/s

71. Problem-solving practice 1

1. 312
2. £3610.80
3. 48 mph

72. Problem-solving practice 2

4. 24.72 litres
5. 3.6 litres
6. 25 cm

GEOMETRY & MEASURES

73. Angle properties

$x = 59°$

74. Solving angle problems

$\angle BDC = x$ (Base angles of an isosceles triangle are equal)
$\angle DBC = 180 - 2x$ (Angles in a triangle add up to 180°)
$\angle ABC = x$ (Base angles of an isosceles triangle are equal)
$\angle ABD = x - (180 - 2x) = 3x - 180$

75. Angles in polygons

1. 30
2. 5040°

76. Pythagoras' theorem

(a) $y = 3.5$ cm   (b) $z = 9.1$ cm

77. Trigonometry 1

(a) 44.8°   (b) 56.2°   (c) 48.6°

78. Trigonometry 2

(a) 4.4 cm   (b) 5.4 cm   (c) 15.7 cm

79. Solving trigonometry problems

4.8 m

80. Perimeter and area

180 cm²

81. Units of area and volume

1. (a) 23 000 cm²   (b) 0.4 cm³
2. 3.5×10^8 mm³
3. 37 500 N/m²

82. Prisms

(a) 600 cm³   (b) 660 cm²

83. Circles and cylinders

119 cm²

84. Sectors of circles

12.6 cm (3 s.f.)

85. Volumes of 3D shapes

806 mm³ (3 s.f.)

86. Surface area

1. 878 cm²
2. 115 cm²

87. Plans and elevations

(a) 9   (b) 16   (c) 9

88. Translations, reflections and rotations

(a) Rotation, 90° clockwise, centre (2, 2)
(b) Translation with vector $\begin{pmatrix} -5 \\ -2 \end{pmatrix}$

89. Enlargement

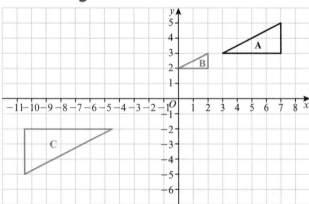

90. Combining transformations

Rotation 180° about point (4, 5)

91. Bearings

(a) 248°   (b) 079°

92. Scale drawings and maps

1. 4.08 km² (3 s.f.)
2. 43 km (accept 42–44 km)

93. Constructions 1

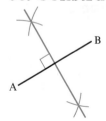

94. Constructions 2

1.

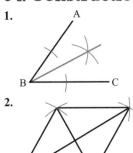

2.

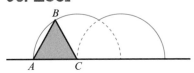

95. Loci

96. Congruent triangles

AC is common to both triangles
AB = AD given
BC = DC given
Triangle ABC is congruent to triangle ADC (using SSS)

97. Similar shapes 1

(a) 130°   (b) 31.5 cm

98. Similar shapes 2

$W = 60$ cm $X = 765$ cm^2 $Y = 290$ cm^3 $Z = 36\,250$ cm^3

99. The sine rule

(a) 45.2° (3 s.f.)   (b) 15.2 cm (3 s.f.)

100. The cosine rule

Angle $PQR = 39.8°$ (3 s.f.)

101. Triangles and segments

Area of sector $= \frac{135}{360} \times \pi \times 4^2 = 6\pi$

Area of triangle $= \frac{1}{2} \times 4 \times 4 \times \sin 135°$

$= \frac{1}{2} \times 4 \times 4 \times \sin 45°$

$= \frac{1}{2} \times 4 \times 4 \times \frac{1}{\sqrt{2}}$

$= 4\sqrt{2}$

Area of segment $= 6\pi - 4\sqrt{2}$

102. Pythagoras in 3D

$DF = \sqrt{20^2 + 25^2 + 10^2} = \sqrt{1125} = 33.541... \text{ cm} > 33 \text{ cm}$
So the needle will fit inside the box.

103. Trigonometry in 3D

27.1° (3 s.f.)

104. Circle facts

63.5°

105. Circle theorems

angle BAC = angle $CBQ = x$ (Alternate segment theorem)
angle $BAD = 180° - 116° = 64°$ (Opposite angles of cyclic
$x + x + 8 = 64$ quad add up to 180°)
$\quad\quad\quad x = 28°$

106. Vectors

(a) $\begin{pmatrix} 6 \\ 10 \end{pmatrix}$ (b) $\begin{pmatrix} 5 \\ -4 \end{pmatrix}$ (c) $\begin{pmatrix} -7 \\ -24 \end{pmatrix}$

107. Vector proof

(a) $AB = -2\mathbf{a} + 4\mathbf{b}$

$\overrightarrow{HA} = -\mathbf{b} + 2\mathbf{a}$ or $2\mathbf{a} - \mathbf{b}$

$\overrightarrow{JS} = 2\mathbf{b} - \frac{2}{3}(-2\mathbf{a} + 4\mathbf{b})$

$= 2\mathbf{b} + \frac{4\mathbf{a}}{3} - \frac{8\mathbf{b}}{3} = \frac{4\mathbf{a}}{3} - \frac{2\mathbf{b}}{3} = \frac{2}{3}(2\mathbf{a} - \mathbf{b})$

$\overrightarrow{KT} = \mathbf{b} - \frac{1}{3}(-2\mathbf{a} + 4\mathbf{b})$

$= \mathbf{b} + \frac{2\mathbf{a}}{3} - \frac{4\mathbf{b}}{3} = \frac{2\mathbf{a}}{3} - \frac{1\mathbf{b}}{3} = \frac{1}{3}(2\mathbf{a} - \mathbf{b})$

$\overrightarrow{HA}$, $\overrightarrow{JS}$ and $\overrightarrow{KT}$ are all multiples of $(2\mathbf{a} - \mathbf{b})$, so are parallel.

(b) $KT : JS : HA = 1 : 2 : 3$

108. Problem-solving practice 1

1. 5.81 m (3 s.f.)
2. 37.5 cm^3

109. Problem-solving practice 2

3. Interior angle of pentagon = 108°
Interior angle of A = 360° − 108° − 90° = 162°
Exterior angle of A = 180° − 162° = 18°
$n = 360° \div 18° = 20$
4. 31.7 cm^2 (3 s.f.)
5.

$\angle OCB = \angle OBC$ (Base angles of isosceles triangle are equal)
$\angle BOC = 180° - 2p$ (Angles in a triangle add up to 180°)
$\angle BOC = 180° - q$ (Angles on a straight line add up to 180°)
So $180° - q = 180° - 2p$
So $q = 2p$
Similarly, $y = 2x$
$\angle BOA = q + y = 2p + 2x = 2(p + x) = 2\angle BCA$

PROBABILITY & STATISTICS

110. Mean, median and mode

19

111. Frequency table averages

0.225 seconds

112. Interquartile range

(a) 19   (b) 5.4 cm   (c) 1.7 cm

113. Line graphs

(a)
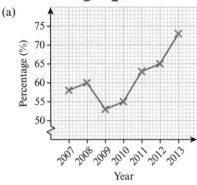

(b) Upward trend (or upward trend from 2009 onwards)

114. Scatter graphs

(a) Accept 23.5° to 24.5°
(b) This is not a reliable estimate because it is extrapolation.

115. Sampling

(a) Amy only observed 6 trains so her statement is unreliable.
(b) $\frac{8}{25} = 0.32$

116. Stratified sampling

(a) 19
(b) e.g. Number each employee and select random numbers using a computer program.

117. Capture-recapture

350

118. Cumulative frequency

Median = 0.225 seconds   IQR = 0.045 seconds

119. Box plots

(a)
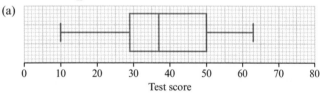

(b) 30 boys   (c) approximately 15 boys

120. Histograms

(a)
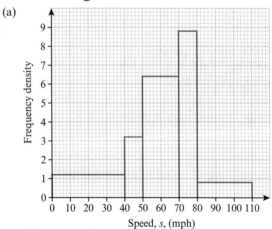

(b) $\dfrac{\left(\frac{128}{4} + 88 + 24\right)}{320} \times 100 = \dfrac{144}{320} \times 100 = 45\%$

121. Frequency polygons

(a)
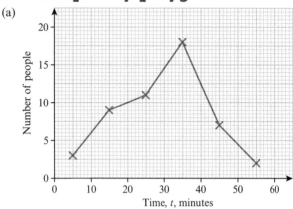

(b) $30 < t \leqslant 40$

122. Comparing data

(a)
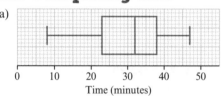

(b) On average, the boys completed the task in the shorter time (lower median).
The girls' times were more consistent (smaller IQR).

123. Probability

0.06 or $\frac{3}{50}$

124. Relative frequency

(a) 0.525 or $\frac{21}{40}$
(b) 0.25 or $\frac{1}{4}$
(c) No. The experimental probability is much greater than the theoretical probability.

125. Venn diagrams

(a) $\frac{5}{30} = \frac{1}{6}$
(b) $\frac{15}{30} = \frac{1}{2}$

126. Conditional probability

1. $\dfrac{1 + 7}{4 + 1 + 7 + 12} = \dfrac{1}{3}$

2. $\dfrac{20}{4 + 17 + 20 + 9} = \dfrac{2}{5}$

127. Tree diagrams

0.789

128. Problem-solving practice 1

1. (a) e.g.

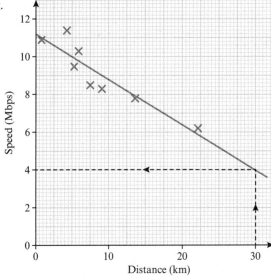

Estimate: 3.4 Mbps

(b) e.g. The prediction is based on a line of best fit/reading from a graph

The reading is extrapolation, 30 km falls outside the range of the data

2. (a) $\frac{7}{64}$ (b) $\frac{49}{64}$

129. Problem-solving practice 2

3. (a) $\frac{1}{6}$ (b) $\frac{1}{3}$

4.

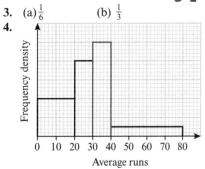

Average runs, x	Frequency
$0 \leqslant x < 20$	4
$20 \leqslant x < 30$	4
$30 \leqslant x < 40$	5
$40 \leqslant x < 80$	2

5. 0.602

Notes

Notes

Notes

Notes

Published by Pearson Education Limited, 80 Strand, London, WC2R 0RL.

www.pearsonschoolsandfecolleges.co.uk

Copies of official specifications for all Edexcel qualifications may be found on the website: www.edexcel.com

Text © Harry Smith and Pearson Education Limited 2015
Copyedited by Andrew Briggs
Typeset and illustrations by Tech-Set Ltd, Gateshead
Original illustrations © Pearson Education Limited 2015
Produced by Out of House Publishing
Cover illustration by Miriam Sturdee

The right of Harry Smith to be identified as author of this work has been asserted by him in accordance with the Copyright, Designs and Patents Act 1988.

First published 2015

18 17 16
10 9 8 7 6 5 4 3

British Library Cataloguing in Publication Data
A catalogue record for this book is available from the British Library

ISBN 978 1 447 98809 0

Printed by L.E.G.O.

A note from the publisher
In order to ensure that this resource offers high-quality support for the associated Pearson qualification, it has been through a review process by the awarding body. This process confirms that this resource fully covers the teaching and learning content of the specification or part of a specification at which it is aimed. It also confirms that it demonstrates an appropriate balance between the development of subject skills, knowledge and understanding, in addition to preparation for assessment.

Endorsement does not cover any guidance on assessment activities or processes (e.g. practice questions or advice on how to answer assessment questions), included in the resource nor does it prescribe any particular approach to the teaching or delivery of a related course.

While the publishers have made every attempt to ensure that advice on the qualification and its assessment is accurate, the official specification and associated assessment guidance materials are the only authoritative source of information and should always be referred to for definitive guidance.

Pearson examiners have not contributed to any sections in this resource relevant to examination papers for which they have responsibility.

Examiners will not use endorsed resources as a source of material for any assessment set by Pearson.

Endorsement of a resource does not mean that the resource is required to achieve this Pearson qualification, nor does it mean that it is the only suitable material available to support the qualification, and any resource lists produced by the awarding body shall include this and other appropriate resources.